THE OXFORD BOOK OF
FLOWERLESS PLANTS

THE OXFORD BOOK OF
FLOWERLESS PLANTS

FERNS, FUNGI, MOSSES AND LIVERWORTS

LICHENS, AND SEAWEEDS

Illustrations by
B. E. NICHOLSON

Text by
FRANK H. BRIGHTMAN

OXFORD UNIVERSITY PRESS
1966

Oxford University Press, Ely House, London W.1

GLASGOW NEW YORK TORONTO MELBOURNE WELLINGTON
CAPE TOWN SALISBURY IBADAN NAIROBI LUSAKA ADDIS ABABA
BOMBAY CALCUTTA MADRAS KARACHI LAHORE DACCA
KUALA LUMPUR HONG KONG

Printed in Great Britain by Jesse Broad & Co. Ltd., Old Trafford, Manchester

Contents

ACKNOWLEDGEMENTS

The 688 plants illustrated in this book have, with the exception of 12 species, been drawn from actual specimens. A very large number of these were collected by the artist and the author themselves. Many of the fungi were found by Mr. J. Hindley, Shaftesbury Modern School, who has given valuable and unstinting help during the whole course of preparing this book.

We would also like to thank Mr. Bruce Ing, warden of Kindrogan Field Centre, Perthshire; Mr. Oliver Gilbert, Dept. of Botany, The University of Newcastle; Mr. E. C. Wallace, Secretary of the British Bryological Society; Simon Wilson; Mrs. A. G. Side; and Mrs. H. M. Howitt, all of whom provided us with specimens. We also thank the East London Diving Club who dived for some of the deep-sea seaweeds at Lulworth.

We acknowledge with gratitude those private people or societies who granted permission to search for plants in private lands: in particular, Mr. Rolf Gardiner, Mr. Michael Pitt Rivers, and The Natural Environment Research Council, Furzebrook, Dorset. Some of the seaweeds were found from Dale Fort Field Centre in Pembrokeshire.

We wish to thank all those who have helped with identifying specimens and checking drawings. These include The Director and Staff, Royal Botanical Gardens, Kew; The Director of the Commonwealth Mycological Institute, Kew; The Dept. of Botany, British Museum, in particular, Mr. J. B. Evans, Mr. P. W. James, Mr. J. R. Laundon, and Mr. J. Price; Mr. A. E. Wade of the Museum of Wales; and T. W. Dunston, Donhead St. Mary, Shaftesbury. We are also grateful to Dr. F. B. Hora, Dept. of Botany, Reading University, who checked most of the Fungi drawings.

The scientific names used throughout the book are, with very few exceptions, those given in the most up-to-date official lists of nomenclature, to be found on p. 202. We appreciate the care and skill of the plate-makers in reproducing the artist's work with such a high degree of accuracy.

INTRODUCTION

This book is a companion book to the *Oxford Book of Wild Flowers*. It is addressed to people of all ages who take pleasure in observing the living and growing things of the countryside. Beyond the world of flowers and trees, but intimately intermingled with it, is another world of flowerless plants. Although many of them are comparatively simple in structure, they are multifarious in form and colour, and they are quite as interesting and fascinating as the more familiar plants. Unfortunately, although flower books abound, those dealing with plants without flowers are few and in general highly technical in their approach. This book is intended to be in some measure an introduction to these scholarly works. If this book helps to bring the unfamiliar plants to people's notice, then some readers at least will pass on to more comprehensive books, a selection of which is listed under 'Further Reading' (page 202).

There are so many different kinds of flowerless plants that only a selection can be mentioned here. They are grouped in the natural localities in which they grow. There are five main sections — seashore, grasslands, uplands, wet places, and woodlands — which have been further subdivided as shown in the list of Contents (page v). Within each subdivision similar plants are grouped together. Five major groups are included — ferns and 'fern allies', mosses and liverworts, fungi, lichens, and seaweeds. A general account of these groups, and their main subdivisions, will be found on page 193.

The primary purpose of this book is to help the beginner with identification, though some information of general interest is given in the descriptions of the plants which appear opposite the colour plates. To know the name of a plant is to hold the key to obtaining further knowledge about it. There are no English names for the majority of the plants described here; most people have been content to speak generally of, say, lichens, seaweeds, or mosses, without distinguishing between the many different species in each group. Each species, however, has a scientific name. This is Latin in form and consists of two words, the first being a sort of surname borne by a group of related species (called a genus), and the second is applicable only to the particular species itself. The correct scientific (Latin) name can be found out by following a set of internationally agreed rules, which are based on the idea that the first person to describe a species is entitled to name it and have the name accepted universally. Everybody need not know the rules nor be able to apply them in detail, any more than it is necessary for a law-abiding citizen to know every detail of the law of the land. In cases of difficulty an expert may be consulted. But by using these names one can be certain of avoiding confusion and misunderstanding. It is possible to invent English names, but if every writer were entitled to follow his own fancy, there would be no way of being certain which plant was being referred to. One plant may have several English names, and one name may be used for more than one plant.

In this book the practice adopted in the standard work on British plants (*Flora of the British Isles*, by Clapham, Tutin and Warburg, 1962, C.U.P.) has been followed. In the few cases where a real English name exists — that is, a name that has really been used, and is not just a book name — it is given; and in other cases where a book name is available that is not merely a direct translation of the scientific name, this is also given but is enclosed in quotation marks. The reader is encouraged to make use of the scientific names, for he will then be able to find his way about in other books, and also make himself understood in other countries. Local names differ from place to place, but scientific names are universal.

No special elaborate apparatus is needed for the elementary study of flowerless plants, but as many of them are small, and the details of their structure are very tiny, a pocket magnifier or hand-lens magnifying between ten and twenty times, which may be purchased for a few shillings, is very useful.

1 **Adiantum capillis-veneris** ('Maidenhair'). The delicate fronds of this fern are closely set in two rows on a creeping stem, about half as thick as a pencil and covered with narrow, pale-brown scales. The small, wedge-shaped leaflets are pale green in colour and translucent, and the fine, dark brown or purplish black shiny stalks are wiry in texture. The spores are produced on special lobes at the tips of the leaflets, which are folded under, thus protecting the spore cases. The plant grows on sheltered limestone cliffs near the sea, invariably protected by an overhang of the rock. It is rare in Britain, being found only in scattered localities in the south west and along the west coast as far north as Westmorland.

2 **Asplenium marinum** ('Sea Spleenwort'). This fern grows in clefts of sea cliffs all round the coasts of Britain, except in the Shetlands and the east coast from south Yorkshire to Sussex. The fronds grow in tufts from a short stock which is densely covered in purplish-brown or brownish-black scales; they are bright green and have a somewhat fleshy texture. The leaflets are very unequal at the base, often being almost lobed on the side towards the tip of the frond, and having a 'cut off' appearance on the other side. The spore cases develop beneath the fronds in oblong groups (sori) on one side (the side towards the tip of the leaflet) of the branch veins, and are protected by an oblong scale.

3 **Roccella fuciformis** ('Orchil') is a rare lichen found on cliffs in a few localities in the West Country and in Wales. The flattened branches grow in tufts which hang downwards, and are bluish grey in colour, usually with a mauvish tinge; they are not shiny. Bluish-white powdery masses of minute reproductive soredia are to be found along the margins. This plant yields a purple dye, but should never be collected for this purpose because of its rarity. The dye is not very permanent. A related species, *R. tinctoria*, which grows on the shores of the Mediterranean, has been used for dyeing since Roman times, and a litmus is still prepared from *R. montagnei* found in Madagascar. Another similar rare lichen, *R. phycopsis*, is found only in a few localities in the West Country, Wales, and west Scotland. It is smaller than *R. fuciformis*, and the branches usually grow upwards and are considerably less flattened.

4 **Ramalina siliquosa** ('Sea Ivory') grows abundantly wherever there are rocks exposed at high-water mark and above. The flattened branches grow in tufts which hang downwards and are pale greenish-grey, smooth, and shining. The spore-producing apothecia appear as raised yellowish-grey discs. This species has been divided into two (*R. cuspidata* and *R. scopulorum*) because there are two kinds of special substance ('lichen acids') produced by the plants, but the differences involved are slight.

5 **Ramalina curnowii** is similar to and equally widespread as *R. siliquosa*, but less common. The branches are much less flattened, narrower, and usually shorter, and markedly blackened at the base.
Other species of *Ramalina* are illustrated on p. 161.

6 **Trichostomum brachydontium** forms broad, rather flat, bright-green cushions on rocks, on wall tops, and sometimes on bare soil, in sheltered places, usually by the sea, and frequently at high-water mark. The enlarged drawing (6A) shows the tongue-shaped leaves with the prominent yellowish-brown nerves protruding at the tip as short, rather blunt points. These become much twisted when dry, but flatten out again immediately this moss is moistened.

7 **Ulota phyllantha.** A moss frequently found in the spray zone just above high-water mark, and grows on trees as well as on rocks, forming rather loose yellow-green cushions. The leaves become much twisted when dry, but when moist are rather long and tapering, with a well-defined nerve extending beyond the tip as a short point (see enlarged drawing 7A). Very small, brown, rod-shaped reproductive structures (gemmae) develop on this point.
The similar species *U. crispa* grows only on trees, and is most often found in upland districts in the north and west of Britain. The leaves taper to a point, but the nerve does not protrude beyond this, and no reproductive gemmae are produced.

8 **Grimmia maritima** grows on rocks all round the coasts of Britain, except in the east and south from Yorkshire to Dorset. This moss is never found inland, though it is often seen on the shores of Scottish lochs in situations where it is regularly submerged at high tide. It does not grow on limestone. The plant forms neat, round, olive-green cushions. The dark-green leaves, which become incurved when dry but only slightly twisted, contract sharply into a long narrow tip. The thick, reddish-brown nerve runs into the tip of the leaf, but does not extend beyond it (8A), and there is no 'hair point', as in other species of *Grimmia* (see p. 74).

1 ADIANTUM CAPILLIS-VENERIS 2 ASPLENIUM MARINUM
3 ROCCELLA FUCIFORMIS 4 RAMALINA SILIQUOSA 5 RAMALINA CURNOWII
6 TRICHOSTOMUM BRACHYDONTIUM 7 ULOTA PHYLLANTHA 8 GRIMMIA MARITIMA

1

1 **Lecanora actophila** forms extensive patches on rocks at high-water mark and above. Each patch consists of a thin yellowish-grey crust on the surface of the rock, and has a well-defined margin and a cracked surface. The spore-producing apothecia are small black discs with a grey rim. The very similar lichen *L. helicopsis* grows in the same situations but is whitish grey with no yellow tinge. The two species have been confused in the past and are referred to in some books together as '*Lecania prosechioides*'.

2 **Lecanora atra** ('Black Shields') is a common lichen on rocks by the sea, but is also found inland on rocks and walls. It forms thick pale-grey crusts with a cracked, rough, lumpy surface, each patch being surrounded by a faint bluish-black line. The apothecia, which were called 'shields' by 18th-century naturalists, have black discs and thick grey rims. When cut open, they show a deep violet colour within.
The less common, though far from rare, similar species, *L. gangaleoides*, grows in the same kinds of situation. It is usually a rather darker grey, and when cut open the apothecia are greenish brown within rather than violet.

3 **Lecanora sordida** consists of rather thick, dull greyish-white crusts surrounded by a thin black line. Although it is patterned with a network of fine cracks, the surface is smooth. The spore-producing apothecia, which have grey powdery discs, are sunk into the crust, not raised above the surface as in other species of *Lecanora*. It is common by the sea, but is also frequently found inland on rocks and walls.

4 **Caloplaca marina** is a lichen found only by the sea on rocks at high-water mark. It is a bright, deep orange colour, and consists of scattered granules and spore-producing apothecia with smooth discs and rims.

Caloplaca thallincolor forms neat rosettes adhering very closely to the rock. The smooth surface is usually marked with fine cracks in the centre, and is delicately lobed at the margins. The whole plant, including the apothecia, is a deep orange colour. It is very similar to the inland species *C. aurantia* (p. 75), but is only found on rocks by the sea.

5 **Anaptychia fusca**. The large number of narrow overlapping lobes form conspicuous cushion-like rosettes on rocks by the sea, and sometimes inland as well. The plant is attached to the stone by numerous fine threads, but unlike the crust-forming species of lichen, it is easy to remove it intact with the point of a knife. It is dark brown in colour, but often appears dark green when thoroughly wet. When dry, it is leathery rather than brittle, and the surface is not shiny, which distinguishes it from species of *Parmelia* (p. 63). The apothecia have black discs with prominent notched margins.

6 **Xanthoria parietina** ('Yellow Scales'), variety **ectanea**. *X. parietina* flourishes especially in places where the atmosphere is charged with mineral salts in the form of minute spray droplets or dust particles, and it is abundant near the sea and on farm buildings. It is yellow or orange in colour, wrinkled in the centre and lobed at the margins, and is attached to the surface on which it grows by fine threads. The variety *ectanea*, which is found only on rocks close to the sea, is deep orange and has a large number of narrow, deeply and finely cut lobes, which frequently overlap one another. The apothecia are usually quite large, with prominent rims. A very similar lichen, *X. aureola*, is illustrated on p. 77.

7 **Pelvetia canaliculata** ('Channelled Wrack') is a very common seaweed growing in olive-brown tufts which become greenish black when dry, and forming a belt at high-water mark on rocky shores all round the coast of Britain. Most of the plants are submerged daily during high spring tides, but during neap tides many may be exposed for days at a time. The seaweed retains water to some extent in channels along the underside of the fronds formed by the inrolling of the margins. In places where there is a great deal of spray, some plants may grow at higher levels still, when very small individuals somewhat resemble the lichen *Lichina pygmaea* (p. 7). The reproductive structures are enclosed in flask-shaped pits in the swollen tips of the branches. Another form of *P. canaliculata* (p. 27) may be found in salt marshes. These seaweeds always contain within them the threads of a fungus, *Mycosphaerella pelvetiae*, and its spore-producing structures may be seen as small dark dots about the size of a pin's head on the surface of the seaweed.

8 **Fucus spiralis** ('Twisted Wrack') is found on all but the most exposed rocky shores, and usually occupies a place immediately beneath *Pelvetia canaliculata*. The branches of the olive-brown fronds are spirally twisted, although they seldom make a complete turn. They have paler swollen tips, looking not unlike soaked sultanas. These contain pits with the reproductive structures inside them, which do not anywhere extend to the extreme rim, thus leaving a sterile margin. As in other species of *Fucus* (p. 9), there are small apertures (cryptostomata) dotted over the surface of the fronds.

1 LECANORA ACTOPHILA   2 LECANORA ATRA   3 LECANORA SORDIDA
4 CALOPLACA MARINA   5 ANAPTYCHIA FUSCA   6 XANTHORIA PARIETINA *var.* ECTANEA
7 PELVETIA CANALICULATA   8 FUCUS SPIRALIS

BETWEEN TIDES: SEAWEEDS

Rocky shores show marked zones in which different species of plants and animals are dominant at different levels on the beach. The brown seaweeds illustrated both on page 3 and on this one are usually found in the sequence in which they are described. The zones of lichens are discussed on page 6.

1 **Ascophyllum nodosum** ('Bladder Wrack', 'Knotted Wrack') grows best on sheltered rocky shores, where it is often the most abundant seaweed; it cannot survive on beaches where there is much wave action. Usually it occupies a zone immediately below *Fucus vesiculosus*, but in more sheltered localities their positions may be reversed. In very sheltered places, such as the heads of some Scottish sea lochs, a free-floating form growing in unattached masses may be a conspicuous feature just off shore. *A. nodosum* is a dark olive green which changes to greenish black on drying; it has long, flat, strap-shaped branches, normally attached by a short stalk to a disc-shaped holdfast. The branches fork repeatedly and produce large, tough air bladders at intervals, but these, unlike the bladders of *Fucus vesiculosus*, cannot easily be 'popped' between the fingers. The branches are notched at intervals, from which, in spring and early summer, short branches grow; these have swollen tips looking rather like greenish-yellow sultanas, in which the reproductive structures are produced. Sometimes the plants bear peculiar short wart-like branches, which are really galls caused by large numbers of a minute worm (*Tylenchus fucicola*).

2 **Polysiphonia lanosa** is a thread-like red seaweed which grows almost always on *Ascophyllum nodosum*, although it is found occasionally on species of *Fucus*, and very rarely on rock. It consists of dark red, repeatedly branched, hair-like threads, at the tips of which the reproductive structures are produced in small swellings. Frequently, small outgrowths, looking like minute red cauliflowers, may be seen scattered all over the plant, which are, in fact, another still smaller red seaweed, *Choreocolax polysiphoniae*. There are about twenty other species of *Polysiphonia* that grow round the coasts of Britain in rock pools at mid-shore level, mostly on stones and shells but sometimes on other seaweeds.

3 **Fucus ceranoides** is a plant of brackish water found in estuaries, lagoons, and land-locked bays, where it grows attached to stones by means of a small conical holdfast. It is a greenish-olive colour and has thin, repeatedly forked fronds with a narrow, well-marked midrib. The fruiting branches grow in fan-shaped groups, and each has a pointed, swollen tip containing numerous flask-shaped cavities holding the reproductive structures.

4 **Fucus vesiculosus** ('Bladder Wrack') grows on the middle shore, usually just below *F. spiralis* (p. 2), but sometimes with *Ascophyllum nodosum* in between. The repeatedly-forked dark olive-brown fronds are joined to a disc-shaped holdfast by a cylindrical stalk. The branches have wavy margins and contain air bladders which are usually in pairs, one on either side of the conspicuous midrib. This is the commonest and most adaptable of all seaweeds, and is very variable. In exposed places plants with very few or no bladders may be found, and very small forms grow in salt marshes (*see* p. 27).

5 **Fucus serratus** ('Serrated Wrack') is very widespread and common, and grows at a low level on the middle shore. The short-stalked, olive-brown fronds branch irregularly, and the margins are jagged with coarse, irregular, forward-projecting teeth. The tips of the branches, where reproductive structures develop in flask-shaped pits, are less swollen than in other species of *Fucus*. There are quite conspicuous minute openings (cryptostomata) scattered over the surface of the fronds. Older plants frequently have the white spiral tubes of a marine worm (*Spirorbis borealis*) adhering to them.

6 **Himanthalia elongata** ('Thong Weed') is to be found attached to rock surfaces below the *Fucus serratus* zone. The young plants look like olive-brown buttons as they grow larger, they become top-shaped, and eventually hollowed on the upper surface like a funnel. From the centre grow strap-shaped branches which fork a few times and hang downwards. The reproductive structures develop on these in scattered pits.

7 **Dictyota dichotoma** is quite common in the south and west, but much less so on other coasts; it grows in rock pools on the lower part of the shore, and also in deeper water. The yellowish or olive-brown plant is rather delicate and translucent; it is sometimes quite pale, and the surface is iridescent. The fronds, which have no midrib, fork repeatedly in a regular manner and the tips of the branches are rounded and blunt. The reproductive structures appear as very small brown spots scattered over the surface.

1 Ascophyllum Nodosum   2 Polysiphonia Lanosa   3 Fucus Ceranoides   4 Fucus Vesiculosus
5 Fucus Serratus   6 Himanthalia Elongata   7 Dictyota Dichotoma

HALF LIFE SIZE

The part of rocky shores where barnacles are to be found serves as a useful reference in describing where on the beach other animals and plants may live. Broadly speaking, the dominant plants higher on the beach than the barnacles are lichens; and brown seaweeds (p. 4) are found lower than the barnacles. Immediately above the barnacles there is a black zone dominated by *Verrucaria maura*. Above this there is an orange zone with *Xanthoria parietina* and various species of *Caloplaca*, and higher still is a grey zone with several different *Lecanora* species (*see* p. 2).

1 **Verrucaria maura** is a lichen forming very extensive dull black crusts on rocks by the sea all round the coasts of Britain. The surface is covered with a network of very fine cracks. The spores are produced in small flask-shaped structures (perithecia) which are sunk in the crust, the openings of which may be seen as minute raised dots.

2 **Verrucaria mucosa** is found wherever there are rocks between tide marks at mid-shore level. The crusts, though extensive, are usually less continuous than those of *V. maura* and also somewhat thicker; they are very dark green rather than black, the colour becoming paler on drying. The surface is quite smooth, with no cracks. The spore-producing perithecia are minute and sunk deeply in the crust, only their tiny openings being visible.

Arthopyrenia halodytes is a minute lichen very commonly found growing on the protective plates of barnacles, with the spore-producing perithecia immersed in tiny pits in the shell.

3 **Lichina pygmaea** is found on the middle shore, forming small, dark-brown densely-branching tufts in crevices in the rock. The branches are somewhat flattened, and the spores are produced in the globular swollen tips of some of the branches. A similar, equally common species, but only half the size and growing at a higher level on the shore, is *L. confinis*. The branches are cylindrical, not flattened.

Catenella repens is a red seaweed, similar in size to *Lichina pygmaea* but dark purple or almost black in colour. It grows in cracks in the rock at mid-shore level and above. The branches are swollen and jointed, making the plant somewhat resemble an extremely miniature form of a succulent plant such as spineless prickly pear.

4 **Laurencia pinnatifida** ('Pepper Dulse'). This seaweed sometimes forms dense, short, crowded tufts, which remain small and never reach maturity, in crevices in rock faces on the middle shore. It is then usually yellowish green in colour, like the fully-grown specimen from a sunlit shallow pool shown here (4A). Plants growing further down the beach and in shade are a dark reddish-purple colour (4B). The holdfast is disc-shaped, but root-like structures also develop from the base of the stalk and not only provide extra anchorage but also may produce new fronds. The main stem and its branches, which are alternately arranged and further divided and subdivided, are much flattened. Spores are produced in summer in tiny oval structures which form dark masses at the tips of the branches. The plant has a strong, distinctive smell and a peppery taste. Other species of *Laurencia* are illustrated on p. 9.

5 **Lithothamnion lenormandi** is a red seaweed which forms quite extensive patches of a hard, reddish-violet crust on the surface of rocks. It flourishes in shade and is common in rock clefts or on boulders covered by the fronds of *Fucus* species; but it may also be found on stones or shells. The crust is thin, with a slightly thickened pinkish-white margin, and the surface is often marked with alternating lighter and darker zones or very thin concentric white lines. There are eleven other species of *Lithothamnion* that are known to grow on the coasts of Britain, but these cannot be distinguished with certainty without detailed microscopical examination. Before making such an examination, it is necessary to dissolve away the calcium carbonate which makes up the bulk of the crust. Some species have thick crusts with wart-like outgrowths on the surface, which make them easy to confuse with the next species.

6 **Lithophyllum incrustans** forms thick crusts on rocks, is pinkish-violet or mauve in colour, and has a knobbly surface. It is common, especially in sunny pools on the middle shore. There are eight other species in Britain, not easy to distinguish from each other.

7 **Hildenbrandia prototypus** consists of dark red patches which lose their lustre on drying, and can very easily be flaked off from the rock. It is common on shaded rocks at mid-shore level. *H. crouani*, which is brownish red rather than dark crimson, is the only other species and is rare.

Rhodochorton purpureum forms crimson, velvet-like patches on rocks on the middle shore, and also on gravelly beaches where the tangle of hair-like filaments becomes matted with sand. The only other species *R. floridulum*, also grows amongst sand but forms dense, globose tufts of dark red threads.

1 Verrucaria Maura

2 Verrucaria Mucosa   3 Lichina Pygmaea   4 Laurencia Pinnatifida

5 Lithothamnion Lenormandi   6 Lithophyllum Incrustans   7 Hildenbrandia Prototypus

1 **Furcellaria fastigiata** grows attached to rocks by means of branching, root-like structures, and the dark purplish-red cylindrical stems are forked six to eight times. The spores are produced in summer and develop in elongated pod-like swellings at the tips of the branches, which fall off at the end of the season after the spores have been dispersed. It is common not only in rock pools between the tides, but also below low-water mark to considerable depths.

2 **Gigartina stellata** is found on the middle and lower shore, frequently very abundantly, especially on the west coast. The flat fronds are dark brownish red, forked six or seven times, and attached to stones or rock by a disc. The margins of the fronds, at least in the lower parts, are inrolled to form a groove or channel, somewhat as in the brown seaweed *Pelvetia canaliculata* (p. 3). In summer and autumn small outgrowths, shaped rather like grape pips, develop on the surface, and spores are produced within them. Old plants are frequently partly overgrown by a brownish-grey layer consisting of colonies of a minute marine animal, the 'Sea-Mat' (*Flustrella hispida*). *G. stellata* has been harvested commercially to produce agar for bacteriological purposes; also it is edible, like Carragheen (No. 6).

3 **Ulva lactuca** ('Sea Lettuce') is a green seaweed found at all times of the year on all types of shore, attached to stones or rock. Plants which have come adrift may continue to flourish in a free-floating state in sheltered situations. The fronds are irregular in outline, wavy, especially at the margins, and translucent. It will grow in places where fresh-water runs into the sea and also where there is a moderate amount of pollution. Old plants often become covered with small brown patches formed by the filaments of the brown seaweed *Myrionema strangulans*, which grows on the surface of the frond. *U. lactuca* has been eaten as an inferior substitute for Laver (*see Porphyria umbilicalis*, p. 23).

4 **Monostroma grevillei** is a rather short-lived green seaweed found on the lower shore in early spring and summer. It develops, attached to rock, as a delicate, transparent bladder which eventually splits and opens out to form a somewhat irregular sheet, very much like *Ulva lactuca*. The frond is, however, thinner, although this is difficult to see except with a microscope, and it adheres firmly to paper when pressed and dried, which *U. lactuca* does not.

5 **Laurencia obtusa** is pinkish purple in colour, sometimes with a yellowish tinge when growing in exposed situations, and is to be found in summer only. It has a cylindrical stem, and the stiff branches, mostly arranged in opposite pairs, branch again twice or three times. It is similar in appearance to *Chondria dasyphylla* (p. 25), though the branchlets of the latter are sharply narrowed towards the base. *L. obtusa* frequently grows upon other seaweeds.

Laurencia hybrida is a dark purple colour and grows on rocks and shells. The branches and branchlets are arranged alternately. *L. pinnatifida*, which has flattened branches, is illustrated on p. 7.

6 **Chondrus crispus** ('Carragheen') grows on all types of shore where there are rocks or stones. The fronds, which grow in clusters, have either a distinct flat stalk, repeatedly forked six to eight times, or a very short stalk when the fronds are broadly fan-shaped. They are usually purplish red in colour but may turn green in strong sunlight. Some forms resemble *Gigartina stellata*, but the margins are never inrolled. Carragheen is one of the most widely used edible seaweeds, and used to be thought to have valuable medicinal properties. The plants are washed in fresh water and then boiled. Weak infusions, with the addition of suitable flavouring, may be used as a beverage, and more concentrated extracts, with sugar and fruit juice added, as table jellies.

7 **Padina pavonia** ('Peacock's Tail') grows in summer and autumn at mid-shore level on sunny beaches in southern England. The fan-shaped fronds, which grow in clusters, are banded on one side with alternating lighter and darker zones of various shades of brown, often with an olive-green tinge; on the other side there is a thin powdery coating of chalk. The margins of the fronds and the edges of the zones are bordered by lines of very short, fine, brown hairs.

8 **Polyides rotundus** ('Goat Tang') is attached to rocks by means of a stout disc, and has dull red cylindrical stems which are forked five or six times. It is very like *Furcellaria fastigiata* in appearance and also may extend from mid-shore down to considerable depths. It differs however, both by the manner of attachment and by the way the spores are produced in oval wart-like growths on the stems, which develop in winter. Also, when the plant is held up to the light it is distinctly red in colour, whereas *F. fastigiata* is blackish brown.

1 Furcellaria Fastigiata   2 Gigartina Stellata
3 Ulva Lactuca   4 Monostroma Grevillei   5 Laurencia Obtusa
6 Chondrus Crispus   7 Padina Pavonia   8 Polyides Rotundus

1 **Gracilaria verrucosa** grows on rocky beaches and gravelly flats at mid-shore level. It is attached to rocks and stones by a small, fleshy disc-shaped holdfast, which may be exposed or buried under sand. The fronds are extensively and very irregularly branched, the branches ending in smooth, slender points. It is a dull, purplish red and has a stringy texture, especially when dry. The spores are produced in summer and early autumn in globular structures scattered all over the plant. *G. verrucosa* is found on suitable shores all over the world, and in southern Australia it has been harvested to produce agar for bacteriological purposes.

2 **Gastroclonium ovatum** is found amongst other seaweeds in rock pools on the middle and lower shore, attached to the rock by a small disc-shaped holdfast which has root-like outgrowths at the margin. The dark purplish or brownish-red stem has several branches, from the upper parts of which come paler coloured, somewhat flattened, oval bladders which have been called 'leaves' and compared with red grains of wheat. They either grow directly on the branches or have short stalks, and spores are produced in minute globular structures at their tips.

Halopitys incurvus grows in thick dull-red tufts on rocks and stones on the lower shore. It has a tough, cylindrical stem, which becomes rigid on drying, and is branched repeatedly, the final branches being directed to one side in a comb-like manner. It is a Mediterranean species, found in Britain only on the south coast.

3 **Dumontia incrassata** is found in rock pools on the middle shore, attached by a very small disc to rocks and stones and sometimes to other seaweeds. In the shade it is dull red, but in full sunlight it appears brownish or yellowish green. The irregularly-branched cylindrical stems have a jelly-like texture and become hollow when fully grown. The branches are narrowest at their point of attachment and have blunt round ends. The plant is common all round the coasts of Britain in the spring, but dies away in late summer.

Nemalion helminthoides is similar to *D. incrassata* and has brownish-purple, somewhat worm-like, slimy branches. However, the branches are more regularly forked and do not become hollow. It is a short-lived

plant, only to be found in late summer, growing on rocks and shells at mid-shore level, and often flourishing in exposed situations.

Helminthora divaricata has a pale-red or reddish-brown stem, with a jelly-like texture; numerous branches grow out at right angles and themselves branch repeatedly. It is found only in summer, growing on stones or on other seaweeds on the lower shore.

Scinaia furcellata is pink or brownish red and slimy to the touch. The frond is fan-shaped and very regularly forked six or eight times; the branches are somewhat flattened. It is a summer plant, found only on the south-west coasts of Britain.

4 **Griffithsia flosculosa** forms delicate tufts of repeatedly forked hair-like threads, which grow in rock pools on the lower shore, especially on more exposed coasts. The specimen illustrated here is a particularly fine and robust one; it is often not so large. The plant is a bright crimson colour, and loses its colour at once if placed in fresh water. The spores are produced in tiny globular structures with short stalks scattered all over the plant. An uncommon species, *G. corallinoides*, has thicker filaments which are coated with a layer of jelly, giving the plant a beaded appearance in its upper parts and causing it to sparkle in bright light. It is bright crimson in colour and has a strong, distinctive smell.

5 **Lomentaria clavellosa** grows on rocks and sometimes on other seaweeds in rock pools on the lower shore. Several main stems, which are hollow, usually grow from a single small disc-shaped holdfast. There are many branches, similar in structure, from which come smaller thread-like branches, which branch again. All the branches are narrowed at both ends and decrease in length successively, the final ones being sharply pointed. The spores are produced in small urn-shaped structures scattered all over the fronds. The whole plant is bright pink or pinkish red and is limp and rather jelly-like to the touch. It is not unlike small and less well-grown forms of *Chylocladia verticillata* (p. 15) but in that seaweed the hollow in the main stem is divided into compartments by cross partitions, and the spore-producing structures are globular in shape. A related species, *L. articulata*, is illustrated on p. 25.

1 Gracilaria Verrucosa   2 Gastroclonium Ovatum   3 Dumontia Incrassata
4 Griffithsia Flosculosa   5 Lomentaria Clavellosa

1　**Bifurcaria bifurcata** is found only on the south and south-west coasts of Britain on exposed beaches, growing in large rock pools on the middle shore in situations where it is never left uncovered. It is a yellowish-olive colour, but becomes very dark brown or almost black and also very brittle on drying. It has a creeping base, attached to the rock by small discs, from which grow fronds of regularly-forked cylindrical branches. There are usually, but not always, a few small bladders in the stems. In spring and summer the branches bear long, swollen tips containing the reproductive structures, which become broken off in the autumn.

2　**Spongomorpha aeruginosa,** shown here growing on a plant of *Laurencia pinnatifida* (p. 7), is a green seaweed, forming globular tufts of fine branching filaments on other plants in spring and early summer. The threads are much interwoven and matted at the base, giving the tufts a spongy texture. Sometimes separate tufts run into one another and form a conspicuous fringe along the frond on which they are growing.

3　**Cladostephus verticillatus** is a brown seaweed growing attached to rock by means of a small disc holdfast in rock pools on the middle shore. The main stem is thin and wiry and branches repeatedly and rather irregularly, both branches and stems being clothed with whorls of short, hooked branchlets. The whole plant is a dark greenish brown and feels harsh to the touch.

4　**Cystoseira tamariscifolia** grows in pools on rocky and gravelly beaches on the south and west coasts of Britain, where it appears to be increasing. It has a densely shrubby appearance because the whole plant is covered with short spiny outgrowths which are sometimes called 'leaves'. It is olive brown in colour, but shows a beautiful blue and green iridescence under water; when dry it turns almost black. The main stem grows from a disc-shaped holdfast and branches profusely. Small air bladders develop near the ends of the branches, which have swollen oval tips containing the reproductive structures.
A related species, *C. baccata*, has large oval air bladders, often more than twice as large as those of *C. tamariscifolia*, and the spiny 'leaves' are restricted to the reproductive parts.

Halidrys siliquosa is a common seaweed of rocky shores, rather like *Cystoseira baccata* but with stiff, somewhat flattened, regularly branched, olive-brown fronds. The long, pointed, oval air bladders at the tips of the branches are divided by cross partitions sometimes into as many as a dozen compartments.

5　**Pilayella littoralis** is found in spring and summer growing on other brown seaweeds. It forms densely-tufted masses of branching filaments which are olive or dull reddish brown in colour. Several generations are produced, growing at successively lower levels on the beach as the weather becomes warmer; the earliest consists of small plants growing on *Ascophyllum nodosum*, and the later and larger plants on *Fucus vesiculosus*, and then, as shown here, on *F. serratus* (p. 5).

6　**Litosiphon pusillus** appears in summer on *Chorda filum* (p. 19). It consists of long, soft, unbranched filaments, which are greenish yellow when young but become brownish olive with age. They grow in dense tufts which are slimy to the touch.

7　**Ectocarpus arctus** grows on various brown seaweeds or on rocks and is shown here on a plant of *Fucus serratus* (p. 5). It forms yellowish or pale olive-green tufts, which are somewhat jelly-like to the touch. Each tuft consists of very numerous and extremely fine, repeatedly branched filaments.

Spongonema tomentosum grows on other brown seaweeds, or occasionally on rocks and stones, on the lower shore. It forms pale olive-green to rusty brown elongated tufts of long filaments which are rather irregularly branched and matted into 'ropes' usually about the diameter of a pencil at the base, but sometimes twice as thick.

8　**Sphacelaria pennata** forms small, dense-brown tufts on a wide variety of other seaweeds (it is shown here on *Cystoseira tamariscifolia*). The filaments are very stiff. They are branched, and the branches bear large numbers of fine branchlets arranged in opposite pairs, giving a feathery appearance.

1 Bifurcaria Bifurcata   2 Spongomorpha Aeruginosa   3 Cladostephus Verticillatus

4 Cystoseira Tamariscifolia

5 Pilayella Littoralis   6 Litosiphon Pusillus   7 Ectocarpus Arctus   8 Sphacelaria Pennata

1 **Codium fragile** grows in pools on the middle and lower shore all round Britain, though most abundantly on the south coast. It is attached to rock by means of an irregular, somewhat lobed, spongy disc formed of closely-woven filaments. The soft, rather yellowish-green, cylindrical stem forks regularly up to eight or nine times. The whole plant is covered with a dense coat of short, fine hairs, which gives the surface a felty texture. In the south and west it is the preferred food of a beautiful sea slug, *Elysia viridis*, which is dark green in colour with a pattern of pale yellow spots; but as it is the most bulky and succulent green seaweed to be found anywhere round Britain, it is no doubt eaten by many other molluscs wherever it grows. A very similar species, *C. tomentosum*, is a darker green colour, and the stems and branches are only about half as thick.

2 **Chylocladia verticillata** is found in late spring and summer attached to stones or other seaweeds in rock pools at mid-shore level and below. It is usually a pinkish-purple colour, but when fully exposed to the sun it may become yellowish green. The main stem, which grows from a very small disc-shaped holdfast, is hollow, and constricted at intervals which get smaller towards the tip, making it look like a string of graded, rather long, oval beads. There is a cross partition in the stem at each constriction, and a whorl of branches at each constriction, each similarly constricted and branched. Scattered, small, globular spore-producing structures develop all over the frond.

Champia parvula is less than half the size of *Chylocladia verticillata*, and its dull red stems and branches consist of large numbers of very short sections. It is an uncommon plant, found only on the south and west coasts.

3 **Ahnfeltia plicata** forms dense, rigid tufts of tangled, wiry threads which are attached to rocks and stones on the middle shore, especially on exposed beaches, by a thin violet crust-like holdfast about the size of a sixpence. The plant is a very dark purple colour which becomes black on drying, although in bright sunlight the upper branches may be dark green. The wiry texture has given rise to various local names, such as '*Fil de fer*' (France), and 'Landlady's Wig' (Pembrokeshire).

4 **Ceramium rubrum** grows abundantly in pools on the middle shore, attached to rocks, stones, shells, and other seaweeds. In the shade it is deep red in colour, but with increasing light it may become brownish red or even greenish yellow. The filaments branch regularly and profusely, and are usually banded across with alternate light and dark sections. The extreme tips of the branches are divided into two points curved towards one another like a pair of pincers. There are twenty-four other species of *Ceramium* growing on the coasts of Britain, but to distinguish them it is necessary to use a microscope.

5 **Cladophora rupestris** is a common green seaweed forming dense tufts on rocks and often growing beneath the fronds of the large brown seaweeds (p. 5). The filaments are much branched, rather dull green in colour, and harsh and somewhat wiry to the touch. It provides food for a minute brownish-green sea slug, *Limaponta capitata*, about the size of a grape pip, which is often found on the plant in large numbers.

Cladophora pellucida is a bright, clear green, rather rigid, but delicate plant, in which tufts of branches grow from short stalks which are, in fact, greatly elongated single cells. There are several other species of *Cladophora* in Britain, some of which grow in fresh water.

6 **Corallina officinalis** grows just below the water surface at the margins of rock pools on the middle shore, or carpets wet rocks in shady places. The holdfast is a hard crust on the surface of the rock, the size of a sixpence or less, from which grow several main stems with opposite branches which branch again at least twice more. The stems and branches are divided into bead-like segments only a little longer than they are wide, which are covered with a hard deposit of calcium carbonate. The colour varies from dull purple in the shade to yellowish pink in bright light, or white on full exposure to the sun. The spores are produced in urn-shaped structures usually growing at the tips of the branches. Sometimes a related red seaweed, *Choreonema thuretii*, grows as a parasite within the filaments of *C. officinalis*, and its smaller, globose, and somewhat translucent spore-producing structures may be seen on the surface of the plant.

7 **Jania rubens** usually grows in very dense tufts on other seaweeds, especially *Cladostephus verticillatus* (p. 12), on the middle and lower shore. The fine, rose-pink threads fork repeatedly, and are divided into segments four to six times as long as they are wide, and encrusted with calcium carbonate. The plant, which is commonest on the south coast, is sometimes parasitized by *Choreonema thuretii* in the same way as *Corallina officinalis*.

1 Codium Fragile　　2 Chylocladia Verticillata　　3 Ahnfeltia Plicata

4 Ceramium Rubrum　　5 Cladophora Rupestris　　6 Corallina Officinalis　　7 Jania Rubens

15

1 **Eudesme virescens** grows on sand-covered rocks in pools at mid-shore level. It may be found during the summer all round Britain, but it is especially abundant on the south coast. It is yellow or olive brown, limp, and slimy to the touch. The main stem, which is attached to rock by a tiny disc-shaped holdfast, has alternately arranged branches bearing short, blunt branchlets at irregular intervals.

Mesogloia vermiculata is similar to *Eudesme virescens* and grows in summer in similar situations, but it is most abundant on north-east coasts. It is a rather darker yellow brown and slimy to the touch, and the irregularly-arranged branches and branchlets are narrowed at the base and the tip and are uneven in diameter throughout their length.

Sauvageaugloia griffithsiana has irregularly-branched, olive-green fronds which are slimy to the touch, and the branches, which may become partly hollow with age, are covered with fine hairs. This species is widely distributed and quite common.

2 **Leathesia difformis** forms olive-brown, rounded, lobed masses about the size of a plum, usually attached to other seaweeds, but sometimes to rocks, on the middle and lower shore from spring until early autumn. It is solid when young, but becomes hollow with age, and has thick, wavy walls with a rubbery texture and a shiny surface.

Colpomenia peregrina is a thin-walled, hollow sphere, varying in size from a small plum to an orange. It is yellowish green when young, becoming olive brown with age, and is covered with scattered dark-brown spots in which spores are produced. The surface is smooth, but not shiny. It is a Mediterranean species, which first appeared on the south coast of Britain at the beginning of this century. It is regarded as a pest of oyster beds in France, where the plants attach themselves to young oysters. They may become full of air when exposed at low tide and then, when the tide comes in again, they act as floats, and the oysters are carried away by the current.

3 **Asperococcus turneri.** The hollow, unbranched, cylindrical fronds grow in tufts from a small disc-shaped holdfast. Each frond has a distinct stalk, from which it widens abruptly; it is green when small but becomes pale olive brown with age; the walls are so thin as to appear translucent. It is to be found in summer on the middle and lower shore, especially in quiet water, growing attached to stones or larger seaweeds.

Asperococcus fistulosus. The hollow fronds have rather thicker walls than those of *A. turneri*, and they widen gradually upwards from the holdfast, with no distinct separate stalk. They are unbranched, and taper at the ends to blunt points, and they have several shallow constrictions at irregular intervals along their length. *A. fistulosus* is found in summer, attached to stones in pools at mid-shore level.

4 **Scytosiphon lomentarius** grows attached to stones and shells in rock pools on the middle shore. The fronds are hollow, cylindrical, and unbranched, and are joined by short stalks to a disc-shaped holdfast from which they grow in tufts. They are olive brown, shiny and rather slimy to the touch, and have a series of deep constrictions along their length. The plant is most common in the winter; several generations are produced in the year, growing at successively lower levels on the beach as the weather becomes warmer.

5 **Chordaria flagelliformis** has dark brown, rather irregularly branched fronds which grow from a disc-shaped holdfast. The main stem and branches are fringed with minute jelly-like hairs, which make the plant slimy to the touch. It is found in late summer and autumn on rocks in pools at mid-shore level.

Sphaerotrichia divaricata is a rare plant with brown, branched fronds which are jelly-like to the touch. The branches are bunched together, and they become hollow towards the ends in older plants.

1 Eudesme Virescens   2 Leathesia Difformis   3 Asperococcus Turneri
4 Scytosiphon Lomentarius   5 Chordaria Flagelliformis

1 **Chorda filum** ('Sea Lace') grows in spring and summer, forming long, olive-brown, slippery 'cords' about as thick as a pencil, attached to stones and rocks by a disc-shaped holdfast. Young plants (1A) are covered with very fine hairs. When full grown the 'cords' are hollow and extremely tough, due partly to the spiral construction of the walls which makes it almost impossible to split the tube lengthwise. The plant cannot survive exposure to air for any length of time. It frequently becomes overgrown by various smaller seaweeds, including *Litosiphon pusillus* and *Sphacelaria pennata* (p. 13).

Punctaria tenuissima forms tufts of narrow ribbons, pale brown in colour and so thin as to be almost transparent, growing on *Chorda filum* and sometimes on other seaweeds.

Punctaria plantaginea has pointed, oval fronds, wider than those of *P. tenuissima*. It is olive brown in colour, and soft and jelly-like to the touch; the surface is covered with scattered darker dots, which contain the spore-producing structures. It grows in spring and early summer on rocks or on other seaweeds in pools on the middle shore.

Pentalonia fascia is similar to *Punctaria plantaginea* but it is smaller, and there are no dark spots on the fronds. It grows in late autumn and winter on rocks in sandy pools on the middle and lower shore.

2 **Saccorhiza polyschides** ('Furbelows') is the largest seaweed found on the shores of Britain, although it only lives for one year. It grows at the low-water mark of spring tides and below, and is much commoner on the south and west coasts than elsewhere. The holdfast is a massive, hollow, knobbly structure. The fan-shaped frond grows from a flattened, twisted stalk with broad, wavy frills at the margins, and is divided into broad ribbons. Minute apertures (cryptostomata) are scattered over the surface of the plant. These are not present in *Laminaria* species.

3 **Laminaria hyperborea** ('Oar Weed') is usually to be found at a slightly lower level on the shore than *L. digitata*. The holdfast is a tangle of thick, root-like structures, curving downwards on to the rock. The stalk is cylindrical, stiff, and hard, with a rough surface on which other seaweeds such as *Rhodymenia palmata* (No. 6) and *Lomentaria articulata* (p. 25) frequently grow. The frond is oar-shaped rather than fan-shaped and splits lengthwise into ribbons. Each

year it grows up again from the base, and eventually the old upper parts are shed.

Laminaria digitata grows at low-water mark on rocky shores all round the coast of Britain. Considerable areas of it are exposed to the air on gently sloping beaches for a short period during low spring tides. The holdfast consists of thin, root-like structures, which spread over the surface of the rock. The stalk is cylindrical, but slightly flattened, flexible and smooth, and usually free from any growth of other seaweeds; but as the plants are long lived, the stalks of old specimens may have become rough enough to support a few plants of *Rhodymenia palmata* (No. 6). The frond is oar-shaped and divided into ribbons.

4 **Laminaria saccharina** is often attached to quite small stones on muddy and sandy flats, as well as to boulders on rocky shores, by means of root-like structures which develop in several tiers from the base of the smooth, cylindrical stalk. The yellowish-olive frond is a single, long, broad ribbon with very frilly margins. On drying, a whitish deposit may appear on the surface which is sweet to the taste. This is the seaweed used by amateur (and perhaps also professional) weather forecasters as a rough and ready indicator of the humidity of the atmosphere.

5 **Alaria esculenta** is commonest on very exposed shores, where it takes the place of *Laminaria digitata*. The holdfast is made up of spreading, root-like structures. The yellowish-olive stalk continues as a midrib along the whole length of the frond, which is a yellowish green colour, and thin and easily torn. The spores are produced in the swollen ends of numerous short-stalked ribbon-like outgrowths from the main stalk.

6 **Rhodymenia palmata** ('Dulse') is one of the commonest of red seaweeds. The narrowly triangular frond is dark red with purple tints and usually very much lobed and divided, with several smaller flat triangular branches growing from the margins and often dividing again. It grows directly from a disc-shaped holdfast on various large brown seaweeds and on rocks, at low-water mark and also at higher levels on the beach. This plant used to be eaten raw in Scotland and Ireland, but it is very tough and practically flavourless.

Rhodymenia pseudopalmata is rose pink or crimson, with a repeatedly forked frond attached by a stalk to the holdfast. It grows in the same sorts of situations as *R. palmata*, but it is rather smaller and much less common.

1 Chorda Filum   2 Saccorhiza Polyschides   3 Laminaria Hyperborea
4 Laminaria Saccharina   5 Alaria Esculenta   6 Rhodymenia Palmata

QUARTER LIFE SIZE

1 **Calliblepharis ciliata** grows on rocks below low-water mark, attached by a holdfast made up of stout filaments. The short stalk widens into a dark-red flat frond which is pointed at the tip and is sometimes divided in large specimens. Similarly-shaped branches of various sizes grow from the margins of the frond, and the spores are produced in fine-pointed branchlets growing from the edges and also all over the surface of the frond. This seaweed is common on the south coast of Britain but rare further north.

Calliblepharis jubata grows attached to other seaweeds and to rocks on the lower shore on the south and west coasts only. The stalk branches to produce several fronds, which in summer have a fringe of long, tendril-like branchlets on their margins.

2 **Phycodrys rubens** has a small disc holdfast and a long stalk bearing numerous brownish-crimson fronds, closely resembling leaves. Each 'leaf' has a midrib and veins and is deeply lobed at the margins like an oak leaf, or sometimes has more pointed lobes rather like holly leaves. In the winter much of the delicate parts of the 'leaves' becomes worn away by wave action, leaving only the midrib, but new leafy growths develop in the spring, as shown in the lower left-hand branch in the illustration. At the same time spores are produced on small branchlets which develop from the veins. *P. rubens* usually grows on the stalks of *Laminaria* species (p. 19).

3 **Dilsea carnosa** grows in tufts from small disc-shaped holdfasts attached to permanently submerged rocks. The short, cylindrical stalks expand into thick, flat, rounded fronds which may become cut by wave action. They are dark red and completely opaque. The whole plant is extremely tough and crisp in texture. It has been confused with *Rhodymenia palmata* (p. 19), but the fronds of the latter are rather less thick and somewhat translucent and grow directly from the holdfast, not from stalks. Although *D. carnosa* has been called *D. edulis* it does not appear ever to have been used as food, as *R. palmata* certainly was.

4 **Delesseria sanguinea** has bright crimson 'leaves' with wavy but unlobed edges which are shaped rather like those of sweet chestnut, although the margins are not toothed. The veins are arranged in opposite, forwardly-directed pairs, from a well-defined midrib. The 'leaves' are attached to a branched cylindrical stalk which grows from an irregularly-shaped disc holdfast. In the winter the fronds wear away, except for the

midribs, on which small outgrowths containing the spores and new 'leaflets' develop in the spring, as shown in the smaller plant on the left in the illustration. The plant grows in shade below low-water mark and further down the shore to considerable depths, and is attached to rocks or to large brown seaweeds.

5 **Hypoglossum woodwardii.** The rose-pink, short-stalked, repeatedly-branched fronds grow in small clusters from very small disc holdfasts, usually with minute root-like outgrowths at the margins. The stalk continues as a prominent midrib for the whole length of the frond, and the branches, which are ribbon-like with pointed ends, arise at right angles from it and branch again, also at right angles. There are no lateral veins arising from the midrib. This is mainly a plant of the south coast of Britain, where it grows on the stalks of *Laminaria* species (p. 19), or on permanently submerged rocks.

Apoglossum ruscifolium is a darker rose-red colour than *Hypoglossum woodwardii*, and the branches of the fronds have rounded, not pointed, tips and wavy margins. Numerous very fine, although frequently only faintly visible, veins branch at wide angles from either side of the midrib and run almost parallel to one another.

6 **Membranoptera alata** is found in the same sorts of situation as *Hypoglossum woodwardii*, but is considerably more common. The crimson frond is irregularly forked into numerous branches which fork again. The branchlets are pointed at the tips. The midrib is prominent and conspicuous throughout the frond, but the very fine veins which arise in pairs from it are only faintly visible.

Nitophyllum punctatum grows on other seaweeds in pools on the middle and lower shore. It has a very thin and delicate frond, similar to that of *Porphyria umbilicalis* (p. 23) in texture, but a bright rose-pink colour without any tinge of purple or green. It is roughly triangular in outline, and grows directly from a small disc-shaped holdfast without any stalk. The margins are divided into narrow lobes which fork once or twice. Round spore-producing structures develop scattered all over the surface of the frond. A much rarer species is *N. bonnemaisonii*, which has a cylindrical stalk and a fan-shaped frond deeply divided into branching lobes. There are some short veins arising from the top of the stalk.

1 Callliblepharis ciliata   2 Phycodrys Rubens   3 Dilsea Carnosa
4 Delesseria Sanguinea   5 Hypoglossum Woodwardii   6 Membranoptera Alata

1 **Callophyllis laciniata** grows below low-water mark all round the coasts of Britain. The forked, fan-shaped frond is attached by a small disc-shaped holdfast to stones, or, as shown here, to the stalk of *Laminaria hyperborea* (p. 19). The vivid crimson colour does not fade with age or on drying. Reproductive structures develop on small lobes which are produced in large numbers on the edges of the frond in late summer.

2 **Ptilota plumosa** is to be found on the coasts of Scotland, north England, and west Ireland, usually growing on *Laminaria* stalks. It is dull red in colour and has irregularly-arranged, firm, flattened branches. The branchlets grow in opposite pairs and have closely-set finer divisions, which give them a feather-like appearance.

Plumaria elegans is similar to *Ptilota plumosa* but somewhat smaller, and the main branches are markedly soft and limp. It forms dark crimson tufts hanging from rocks on the middle and lower shore. The plants provide a favourite place of attachment for the common purse sponge, *Grantia compressa*. This looks like a small somewhat flattened vase, yellowish-grey in colour and about the size of the bowl of a teaspoon.

3 **Odonthalia dentata** grows attached to rocks or to *Laminaria* stalks on the coasts of Scotland and north England. The tough flattened fronds are dark red in colour and have a distinct, somewhat peppery, but not unpleasant smell. They are irregularly branched, and have rather short, even branchlets, which are toothed at the tips. The reproductive structures grow in tufts between these teeth in the early spring.

4 **Cryptopleura ramosa** grows attached to large brown seaweeds or sometimes to rocks, in deep water, or at higher levels on the beach in shaded, deep, rock pools. It has a thin, tough, repeatedly and irregularly branched frond, with a well-developed midrib which becomes indistinct towards the tips of the branches. There are branching veins in the lower parts only, which may be difficult to see. The plant is brownish or purplish red and shows a slight iridescence. It is common all round Britain, but especially on the south coast.

5 **Porphyra umbilicalis** (Laver) has a very thin flexible frond, which is irregularly lobed and grows from a minute disc holdfast. The rosy purple colour readily fades to olive green, and becomes dark and also very brittle when dry. It is common, especially on exposed shores, all round Britain, growing at all levels on the beach attached to rocks and stones, particularly in sandy places. It is used as food in south Wales and Ireland, where it is boiled until tender and then fried with a coating of oatmeal in bacon fat. The plant is cultivated in Japan; bundles of bamboo are placed on the bottom offshore, and when the seaweed has established itself, they are transferred to river estuaries, where, in the slightly brackish water, a lusher, softer growth develops.

The very similar species, *P. leucosticta*, can only be distinguished with certainty by examining the reproductive structures under a microscope. Its frond is not usually lobed, and the rosy-purple colour does not fade so readily. *P. amethystea* is a rare plant with stiff violet fronds.

Porphyra miniata which is to be found in spring and summer on the north and west coasts of Britain and is a deep red colour, differs from other species of *Porphyra* in that it is thicker and less transparent. To be certain of the difference, however, it is necessary to use a microscope.

Polyneura gmelinii is similar in texture to *Porphyra miniata* but is a paler, rose-red colour. It has a disc-shaped holdfast and a short cylindrical stalk that expands into a frond almost circular in outline. The margins are wavy or shallowly divided into irregular lobes. Branched, curved veins run from the top of the stalk about half way across the frond. Spores are produced in small oblong structures at or near the margins. In the very similar species, *P. hilliae*, the round spore-producing structures are scattered all over the frond. Both species grow in shady rock pools near low-water mark on the south and west coasts of Britain.

6 **Plocamium vulgare** grows in pools on the lower shore attached to stones and to other seaweeds by a small branching holdfast. The frond is flattened and consists of a tough, narrow, main stalk which branches four or five times in a regular pattern. The final branchlets arise all on the same side, rather like the teeth of a comb. The whole plant is a clear rose-red colour.

7 **Halurus equisetifolius** is found in the summer months on rocks below low-water mark, or in pools at higher levels on the beach, chiefly on the southern coasts of Britain. The main stem is a fine thread which bears similar, irregularly-arranged branches. All the stems have very numerous closely-set whorls of fine forked branchlets, which overlap and make the branches look like tiny bottle brushes when they are taken out of the water.

1 Callophyllis Laciniata   2 Ptilota Plumosa   3 Odonthalia Dentata
4 Cryptopleura Ramosa   5 Porphyra Umbilicalis   6 Plocamium Vulgare
7 Halurus Equisetifolius

1 **Cystoclonium purpureum** grows in summer on rocks and other seaweeds from low-water mark down to considerable depths. The main stem grows from a holdfast consisting of branched, root-like structures. The stem is cylindrical and about twice as thick as the numerous irregularly-arranged branches, which branch again several times. All the branches and branchlets are narrowed at the base and also tapered to very fine points at the tip. Spores are produced in small globular structures on the smaller branchlets. The whole plant is a dull, purplish-red colour.

2 **Brongniartella byssoides** is found in summer at low-water mark and below. The main stems grow from a creeping stem which is attached by short, fine threads to rocks, shells, and other seaweeds, especially *Furcellaria fastigiata* (p. 9). Alternately-arranged branches grow from the main stem and give rise to branchlets, also alternate, which are themselves closely covered with finer divisions. The spores are produced in small oval structures near the ends of the branches. This plant is less common in the north than it is in the south.

3 **Heterosiphonia plumosa** has a disc-shaped holdfast and a cylindrical main stem with irregularly alternate branches, from which grow numerous branchlets with further subdivisions. These are progressively shorter towards the tips, producing a somewhat feather-like outline. The smaller branchlets may have minute oval structures at their bases, containing the spores. The plant is a deep crimson colour and is commonest on the south and west coasts of Britain.

4 **Chondria dasyphylla** is found in summer, growing on rocks and stones on sandy and muddy bottoms. It has a minute disc holdfast and a tough, stout main stem, with alternate branches bearing short branchlets. Reproductive structures develop on small outgrowths from these. The branchlets are narrowed at the base, but blunt and rounded at the tips. There is a small depression at the extreme apex containing a minute tuft of hairs.
C. tenuissima, a similar but considerably less common species, is more slender in all its parts, and the branchlets are pointed at both ends.

5 **Lomentaria articulata** has branching, hollow stems which are constricted at regular intervals into oval bead-like portions that become progressively smaller towards the tips; one or more branches arise from each constriction. Several stems grow from each tiny disc-shaped holdfast, and some of them creep over the surface of the rock and form additional holdfasts and new plants. The spores are produced in small pear-shaped structures on the upper branches. The whole plant is shiny and pinkish-crimson in colour. A related species, *L. clavellosa*, is illustrated on p. 11.

6 **Phyllophora membranifolia** has a very small disc holdfast with several thin, branching, cylindrical stalks, from which grow flat 'fans' of forking, wedge-shaped branches. The spores are produced in small oval structures growing on the main stalks. The plant, which is brownish purple or purplish red in colour, grows especially on steep-sided rocks at low-water mark and below.

Phyllophora crispa frequently grows in clusters of flat, clear red, ribbon-like fronds which branch only at the tips. Each frond has a very short stalk with a small disc holdfast. Spores are produced in small, globular seed-like structures with wrinkled surfaces which develop on the edges of the fronds, and also on tiny stalked leaflets which grow from the surface of the upper part of the plant.

Phyllophora palmettoides has a large, thick, disc holdfast with several cylindrical stalks which gradually flatten and widen into fan-shaped fronds. They are rose red in colour, rather stiff, and repeatedly forked into wedge-shaped branches. This plant is found on the south coast; it is rare in the west, and does not grow at all in the north and east.

Bryopsis plumosa grows on the steep sides of deep pools on the lower shore in spring and summer. The whole plant is bright green. The main stem bears a few branches with crowded branchlets in two rows, diminishing regularly in length towards the tip and thus producing a feathery appearance.
B. hypnoides is similar, but the branchlets grow from all sides of the stems and may branch again, producing a tufted appearance at the ends. It is only common on shores in the extreme west of Britain.

1 CYSTOCLONIUM PURPUREUM   2 BRONGNIARTELLA BYSSOIDES   3 HETEROSIPHONIA PLUMOSA
4 CHONDRIA DASYPHYLLA   5 LOMENTARIA ARTICULATA   6 PHYLLOPHORA MEMBRANIFOLIA

Salt marshes are found on sheltered coasts and by river estuaries, but they do not provide a very stable habitat, or a very favourable one for flowerless plants. The mud is colonized by flowering plants such as Glasswort, Sea Purslane, and the grass *Spartina townsendii*, shown here, in whose shelter fine 'turfs' of the filamentous blue-green seaweed *Calothrix* frequently develop. The jelly-like blobs of another blue-green seaweed, *Rivularia*, are also common. Apart from these, and the plants described below, no other flowerless plants are likely to be found in this habitat.

1 **Fucus vesiculosus** form **muscosa**. *F. vesiculosus* normally grows at mid-shore level on rocky coasts, and has large fronds with air bladders, attached by a holdfast to the rock (p. 5). The form *muscosa*, which consists of very small, dark-brown, irregular fronds, without holdfasts or bladders, grows in mud, where it sometimes forms a dense 'turf' and may make quite extensive patches. There are several other small and distorted-looking salt-marsh forms of this and other species of brown seaweed, which are frequently very difficult to identify, as they rarely produce reproductive structures.

2 **Pelvetia canaliculata** form **libera**. The normal form grows at high-water mark on rocky shores, firmly attached to the rock (p. 3). The form *libera* is found quite unattached, tangled amongst the flowering plants or lying on the mud in salt marshes, often at very high levels where the tide only occasionally reaches. It can usually be distinguished from similar forms of other brown seaweeds by having some trace of inrolling of the margins, which produces a shallow channel on the underside.

3 **Bostrychia scorpioides** is a red seaweed which grows in tangled tufts of wavy threads with irregular branches. The branches bear alternately-arranged branchlets which are covered with fine, short, sharply-pointed, further divisions and coil spirally at the tips. The plant is usually found tangled about the bases of salt-marsh flowering plants, especially Sea Purslane; here it is shown on the grass *Spartina townsendii*. It also grows amongst *Pelvetia canaiiculata* and *Fucus spiralis* (p. 3) in very sheltered situations. It is commonest on the south and west coasts.

4 **Enteromorpha intestinalis** is a green seaweed with long, unbranched, tubular fronds, which develop from a very small disc holdfast, but frequently become detached and form free-floating masses. It varies in size from about the thickness and length of a pencil to three or four times as large, and when fully grown is irregularly inflated and constricted at intervals. It is abundant on salt marshes and in brackish dikes and ditches, and is also found in pools on the upper shore on rocky coasts, especially, but not always, where some fresh water flows into the sea. The plants begin to grow in the early spring and die away in the autumn, when the dead, bleached fronds may be seen in considerable quantity on the beach. It is used as food in China and Japan, though not, as far as we know, in Britain.

5 **Enteromorpha compressa** has long, bright-green, tubular fronds which branch sparingly from near the base. The branches are somewhat flattened but not markedly constricted or inflated; they are narrowed at the base and blunt at the tip. The plant is common on all kinds of beach at mid-shore level, growing attached to rocks and stones; it also penetrates into brackish water, where it may form a 'turf' on the bottoms of shallow pools.

6 **Enteromorpha linza** grows attached by a small disc holdfast to rocks and stones, and also to other seaweeds — frequently *Corallina officinalis* (p. 15) — in pools on the middle and upper shore, especially where the water is slightly brackish. It has a short, cylindrical, hollow stalk, expanding into a bright-green, flat, ribbon-like frond which is usually spirally twisted and often somewhat crimped at the margins.

Enteromorpha torta consists of rather thick, twisted, bright-green strands, in which the internal hollow is almost or completely closed. They are sparingly branched, and form tangled 'mats' on muddy beaches at mid-shore level. At least eight other species of *Enteromorpha* are known to grow round the coasts of Britain.

1 F<small>UCUS</small> V<small>ESICULOSUS</small> *form* M<small>USCOSA</small>   2 P<small>ELVETIA</small> C<small>ANALICULATA</small> *form* L<small>IBERA</small>

3 B<small>OSTRYCHIA</small> S<small>CORPIOIDES</small>

4 E<small>NTEROMORPHA</small> I<small>NTESTINALIS</small>   5 E<small>NTEROMORPHA</small> C<small>OMPRESSA</small>   6 E<small>NTEROMORPHA</small> L<small>INZA</small>

1 **Polytrichum piliferum** ('Hair Moss') forms bluish-green, open 'turfs' of moss on sand dunes and bare places on well-drained soils. On the upper surface of the leaves, which have sheathing bases, are a series of vertical plates running lengthwise, which make them opaque. The plant is distinguished from other species of *Polytrichum* (p. 80) by the greyish-white, stiff bristle which grows from the tip of each leaf. The spore-producing capsule at the end of a stalk is square in cross-section and is covered when young by a loosely-fitting, very hairy 'cap' (calyptra).

2 **Tortula ruraliformis** grows in extensive, rather loose, yellowish-green or golden-brown patches on sand dunes; this moss is rarely found inland. The leaves, which are bent backwards, taper into long, silvery, hair points. The capsule is cylindrical, and grows upright on a reddish-green stalk. When ripe, the lid at the top falls off, and a conspicuous fringe of 'teeth' can be seen surrounding the opening.

3 **Peltigera canina** ('Dog Lichen') is a large, flat, leaf-like lichen, with broad, branching lobes creeping on the surface of the ground. It is soft, flexible, and brownish green when moist, but brittle and papery and whitish grey when dry (3A). The margins are distinctly thinner than the central parts, and are rolled under. The white and felt-like under-surface has a network of raised veins with root-like structures (rhizines) growing from them which anchor the plant to the soil. The veins and rhizines are usually white but may be various shades of brown. The upper surface is finely downy, and the pattern of the veins can be seen because the spaces between them are slightly domed upwards, rather like the leaf of a Savoy cabbage. The spore-producing structures (apothecia) are chesnut-brown oval discs growing on the top surface of upward-pointing narrow lobes.

4 **Peltigera rufescens** has smaller lobes than *P. canina*. It often has a distinctly reddish tinge when moist and is grey or greyish brown when dry (4A). The lobes are crinkled at the margins, which are somewhat thickened, and the upper surfaces are finely downy, but without the 'savoy cabbage' appearance of *P. canina*. The felted under-surface is white, often tinged with brown, and the veins, which tend to run parallel to one another rather than spreading widely, are usually the same colour as the surrounding parts but sometimes darker brown. The spore-producing structures are like those of *P. canina*.

Peltigera spuria is much smaller than *P. canina*, and has upwardly-growing rounded lobes frequently about the size of a sixpence. The upper surface is downy, and in addition it is often covered with patches of granular reproductive structures (soredia).

5 **Peltigera polydactyla** is brownish green or greenish grey when moist, and brownish grey when dry (5A). The upper surface is smooth and shiny and not at all downy. The felted under-surface is white, tinged with brown, and has a network of rather flat veins. The spore-producing apothecia are reddish brown and oblong, and are produced on long narrow lobes, with turned-back margins, which grow vertically upwards.

Peltigera horizontalis is larger than *P. polydactyla* and is found in moist situations, especially in the west of Britain. The upper surface is smooth and glossy, and the lobes are crinkled at the margins. The spore-producing apothecia are round and flat and develop on narrow lobes which do not grow upwards.

6 **Cladonia foliacea** is a lichen which forms loose mats of small 'leaves' (squamules) on sandy, gravelly, and light chalky soils. Each squamule is yellowish green on the upper surface and greenish yellow on the lower, and is deeply divided at the tip. When dry, it tends to curl upwards. Spore-producing apothecia sometimes develop on the rims of small stalked cups, rather like miniature versions of the cups of *Cladonia chlorophaea*, which grow in groups on the surface of the squamules.

7 **Cladonia cornutoradiata** has very small, greyish-green basal 'leaves' or squamules. The main part of the plant consists of long stalks, which produce an irregular cluster of branches or an asymmetric cup with outgrowths from its margins, with some resemblance to the antlers of a deer. The visible outer surface is grey; there is no grey-green outer layer (cortex), such as is found in many species of *Cladonia*.

8 **Cladonia chlorophaea** grows commonly on light soils, usually in moister places. The basal 'leaves' or squamules are small, rather thick, and grey-green with a brownish tinge. From them grow broad and regular cups, which expand from the base without a stalk. They have a grey-green, unbroken, rather warted outer surface, except at the rim and inside the cup where the surface breaks down into a coarse powder (soredia). This species was at one time regarded as a variety of *C. pyxidata*, which differs, however, in being entirely warted and granular, without any soredia. *C. conista* is another less common but similar species in which the soredia form a fine, flour-like rather than a coarse and granular powder.

1 POLYTRICHUM PILIFERUM   2 TORTULA RURALIFORMIS   3 PELTIGERA CANINA

4 PELTIGERA RUFESCENS   5 PELTIGERA POLYDACTYLA   6 CLADONIA FOLIACEA

7 CLADONIA CORNUTORADIATA   8 CLADONIA CHLOROPHAEA

1 **Lepiota procera** (Parasol Mushroom) grows in summer and autumn in grassy places on the edges of woodlands and in clearings. The cap is greyish brown and has a distinct boss (umbo) in the centre; it is covered with large, dull-brown, shaggy scales. The stem is similarly scaly and has a well-developed double ring, above which it is smooth and greyish white. The gills and spores are white. The soft, white flesh is rather dry and has a distinct nutty flavour. It is a useful edible mushroom because of its size and its good flavour and aroma, but it is not very common.

Lepiota rhacodes ('Shaggy Parasol') grows in open, grassy places on moist, rich soil, and frequently appears to be associated with conifers. It has a very scaly cap with a prominent, shining, dark-brown umbo, but the greyish or brownish white stalk is quite smooth. It is good to eat but is not common. More than fifty other species of *Lepiota* have been found in Britain.

2 **Nolanea sericea** is found from late spring until early autumn growing in meadows and lawns. The shining brown cap has a silky texture; it becomes paler on drying, and frequently splits at the margins. The slender stem is sometimes vertically grooved, and is hollow within. The gills are pale brown and rather distantly spaced; the spores are pink. The pale-grey flesh has a strong mealy taste and smell.

3 **Tubaria furfuracea** grows at all times of the year, but especially in the autumn, in fields and also in open places in woodlands, often on sticks and chips of wood. The cap is reddish brown when moist, but becomes paler on drying; it has slightly grooved margins which are covered with mealy or scurfy patches; in fully-grown specimens it is flat. The pale reddish brown stalk is slender and hollow, with a slightly woolly base. The gills and spores are cinnamon brown.

4 **Stropharia aeruginosa** ('Verdigris Agaric') grows amongst grass in fields and woods in summer and autumn. The deep blue-green cap is slimy to the touch. In young specimens there are soft white scales at the margins, but these disappear with age, and also the colour fades to a dull yellowish green. The long blue-green stalk is scaly below the well-developed ring and smooth above. The gills and spores are dark brown. The soft, thin, pale-green flesh smells and tastes rather like radishes, but should not be eaten.

5 **Stropharia semiglobata** is found in pastures from spring until autumn, usually growing on old dung. It has a pale-yellow hemispherical cap which is slimy to the touch; the slender stalk is also slimy below the narrow ring. The gills and spores are dark brown, and the flesh is pale yellowish white with a mealy smell.

Stropharia coronilla grows in grassland and amongst stubble in summer and autumn. It differs from *S. semiglobata* in that the cap is not hemispherical, but flatter, broader, and thicker; and the well-developed ring is distinctly radially grooved on the upper side.

6 **Agaricus campestris** (Field Mushroom) is found in meadows and pastures in late summer and autumn. The white cap has a silky texture; the stem is also white, rather short and thick, and with a narrow ring which usually disappears in fully-developed specimens. The gills are pink when young, but become dark brown and finally purplish black with age. The spores are dark brown. The flesh, which is soft and flushes a pale flesh pink colour when cut, has a pleasant taste and smell. In Britain this species is traditionally regarded as the most desirable of all edible fungi.

7 **Agaricus bisporus** differs from *A. campestris* in that the spores are produced in twos rather than in fours. It grows on old, well-rotted dung, and is associated with manure heaps and roadsides rather than grassland. In the wild form of this species the cap is tinged a pale fawn colour. A white form *albida* is the cultivated mushroom of commerce.

Agaricus xanthoderma is like *A. campestris* but the flesh stains yellow when bruised or cut. It should not be eaten, for it causes unpleasant symptoms, although apparently it is never fatal. Other species of *Agaricus* are shown on p. 13.

8 **Lepista saeva** (Blewits) grows in pastures in autumn and early winter. The flat cap has an incurved margin and varies from a very pale tan colour to greyish buff. The stem is short, thick, and somewhat swollen at the base, and is bluish grey, streaked with fine, pale violet, fibrous scales. The crowded gills are greyish white, sometimes flushed with pink; the spores are pink. The flesh has a pleasant taste and smell and is good to eat. *Lepista nuda*, a similar edible species found in woodlands, is illustrated on p. 135.

1 Lepiota Procera   2 Nolanea Sericea
3 Tubaria Furfuracea   4 Stropharia Aeruginosa   5 Stropharia Semiglobata
6 Agaricus Campestris   7 Agaricus Bisporus   8 Lepista Saeva

1 **Hygrophorus obrusseus** grows in grassland in summer and autumn. The cap is yellow with an orange tinge and slimy to the touch when wet. It is conical when young, but with age the rim turns upwards, and the top becomes flattened. The stalk and gills are similarly coloured, but paler. As in other species of *Hygrophorus*, the gills are thick, rather distantly spaced, and have a waxy texture, and the spores are white.

2 **Hygrophorus pratensis** is found in pastures and meadows from late summer until early winter. The cap is conical when young, but becomes flattened with age, and is not slimy to the touch. The whole plant is brownish yellow or buff in colour, the stalk and gills being paler than the cap.

3 **Hygrophorus coccineus** grows especially in grassy places on the margins of woodlands, from summer until early winter. The cap is usually somewhat bell-shaped, and is bright scarlet in colour; the stalk is red at the top, shading into yellow at the base, and the gills are yellow, flushed with red. The brightness of the colours fades with age.

4 **Hygrophorus puniceus** is found in late summer, autumn, and early winter in fields and on roadside verges. It is nearly twice as large as *H. coccineus*, and is crimson rather than scarlet, fading to yellowish red with age. The stalk is white at the base and has a rather coarsely fibrous surface.

Hygrophorus conicus is scarlet, flushed with orange, and has a yellow stalk and gills. The whole plant fades to yellow with age, and turns black when bruised. It is about the same size as *H. coccineus*. *H. nigrescens* is very similar, but the stalk is white at the base.

5 **Hygrophorus niveus** appears in pastures and meadows in autumn and early winter. The cap is conical when young, but becomes flattened with age; the stalk tapers towards the base. The whole plant is ivory white in colour.

6 **Hygrophorus psittacinus** ('Parrot Toadstool') is found not only in fields but also in open, grassy places in woods, in summer and autumn. When fully developed, the cap is rather flat, with a distinct boss (umbo) in the centre; it is slimy to the touch. It is bright blue-green, fading through reddish tints to pinkish yellow. *H. hypothejus*, a species of pine woods, is illustrated on p. 103.

7 **Mycena fibula** ('Carpet Pin') grows amongst grasses and mosses in damp places in summer and autumn. The hemispherical cap is pale orange buff, and the stalk is pale orange and downy at the base. The pale yellowish-buff gills are arched and run down the stem. The spores are white. As in most species of *Mycena*, and also *Galerina hypnorum* (p. 47), *Coprinus plicatilis* (p. 35), and *Omphalia sphagnicola* (p. 89), the cap is thin, and the gills can be seen from above as radial ridges.

8 **Mycena flavo-alba** is found on lawns and in short grass in fields in late summer and autumn. It has a conical or somewhat bell-shaped cap which is ivory white in colour, usually tinged with yellow, at least in the centre. The gills are white and do not run down the stalk.

9 **Mycena leptocephala** ('*L'Azoteux*') grows amongst grass and dead leaves in late summer and autumn. It has a conical cap and a long, slender stalk, and the whole plant is smoky grey in colour. The gills have whitish grey edges. The fungus has a characteristic and distinct smell, resembling that of nitric acid. *M. alcalina* is a similar species, but it grows on rotting wood. The cap is greyish brown rather than smoky grey, and the odour of nitric acid is less pronounced.

10 **Mycena olivaceo-marginata** often grows in large groups in short turf in pastures and on lawns in summer and autumn. It is a dull yellowish-brown colour, and the cap is distinctly bonnet-shaped. The flesh is thin and rather watery and smells faintly of radishes. Other species of *Mycena*, which grow in woods, are illustrated on pp. 105, 121, 127, and 133.

11 **Psilocybe semilanceata** ('Liberty Cap') is found on roadside verges as well as in grassland in late summer and autumn. The pale greyish-brown cap is bonnet-shaped with a pointed top and a pleated margin which is turned inwards. It has a thin, tough, jelly-like skin, which is somewhat slimy to the touch and can be peeled off. The stalk is rather long, thin, and wavy. The gills are purplish brown when young, and become almost black as the very dark spores ripen.

1 Hygrophorus Obrusseus   2 Hygrophorus Pratensis   3 Hygrophorus Coccineus
4 Hygrophorus Puniceus   5 Hygrophorus Niveus   6 Hygrophorus Psittacinus
7 Mycena Fibula   8 Mycena Flavo-Alba   9 Mycena Leptocephala
10 Mycena Olivaceo-marginata   11 Psilocybe Semilanceata

1   **Marasmius oreades** (Champignon) grows in groups, often forming fairy-rings in grassland, especially where the turf is short. It is common in closely grazed pasture and on lawns, in summer and autumn. The cap is light brown, usually with a pinkish tinge, and becomes markedly paler on drying. The stalk, which is slightly downy at the base, and the rather distantly spaced gills, are pale buff; the spores are white. It resembles *Clitocybe rivulosa*, except for the slight boss (umbo) in the centre of the cap, even in old specimens in which the rim has turned upwards, and for the fact that the gills do not run down the stem. It differs from *Collybia peronata* (p. 133) in the tough and elastic nature of the stem. The flesh is white and has a faint smell of fresh sawdust, and it is good to eat, with a pleasant mushroom taste. It can be dried and stored satisfactorily, and revives well when wetted again. A much smaller species, *M. ramealis*, is illustrated on p. 145.

2   **Clitocybe rivulosa** ('False Champignon') is to be found in groups amongst short grass in late summer and autumn, and may form fairy-rings, as does *Marasmius oreades*, with which it often grows. The cap is pale flesh-coloured, sometimes tinged with brown; and, when young, the surface is dusted with a fine powder, and there are alternating lighter and darker zones. The stalk, which is slightly hairy at the top, is the same colour as the cap. The rather crowded gills are paler and run down the stalk, and this, together with the presence of a slight hollow (umbilicus) in the centre of the cap, distinguishes the plant from *M. oreades*. The spores and flesh are white, and the almost odourless flesh is poisonous.

3   **Clitocybe dealbata** is very similar to *C. rivulosa*, and has sometimes been regarded as merely a variety of it. The cap is finely powdery but not zoned, and the whole plant is dull white, tinged with brown or yellow. The flesh has a mealy smell and is poisonous. It grows in groups in pastures and sometimes as an abundant weed in mushroom beds. Other species of *Clitocybe*, which grow in woodlands, are illustrated on p. 107.

4   **Conocybe tenera** is found amongst grass in fields and also in woods, from late spring until early winter. The yellowish-brown conical cap has a rusty tinge when wet, and the long, similarly-coloured stalk has a finely powdery surface and a small bulb at the base. The crowded gills and the spores are cinnamon brown.

Conocybe lactea has a milk white or cream coloured, steeply conical cap, a long white stalk, and pale brown gills and spores. It is to be found in open, grassy places, and on sand dunes.

5   **Coprinus comatus** ('Shaggy Ink Cap') grows on rich soil from late spring until autumn, and is commonly found in groups in fields and on disturbed ground such as roadsides and rubbish tips. In young specimens the cap is white and covered with shaggy scales, and the hollow stalk has a conspicuous ring, which soon disappears. The gills are white, although they turn pink later. When fully expanded, the cap and the gills beginning at the bottom, gradually dissolve into a black, inky fluid containing the spores. If gathered before it has begun to dissolve, it is good to eat.

Coprinus sterquilinus is similar to *C. comatus*, but is less than half the size. It is an uncommon plant, found growing on dung. The cap is white and covered with curved and rather pointed scales, and the stalk has a well-developed ring.

6   **Coprinus atramentarius** (Ink Cap) is to be found in late summer and autumn in open, grassy places on rich soil, frequently in clusters near trees. The pale brownish-grey cap is scaly in the centre and grooved at the margins. The gills are greyish white when young, but turn black as the spores develop. The white stalk tapers towards the top and has a ring-like mark at the base. The cap and gills eventually dissolve into a black liquid, as in *C. comatus*. It is edible if cooked when young, but should not be eaten with wine, for the plant contains a substance (bis-diethyl-thiocarbomoyl disulphide) which reacts with alcohol to form a product causing nausea. This substance has been used as the basis of a 'cure' for alcoholism.

7   **Coprinus plicatilis** grows singly amongst short grass in fields and on lawns from late spring until autumn. The cap is brown in the centre and grey at the margins. The greyish-white stalk is long, thin, and brittle, and the gills are grey. As in *Mycena fibula* (p. 33) the gills produce corresponding ridges in the very thin cap, and so can be seen from above. Eventually the cap and gills wither, but do not dissolve as do most other species of *Coprinus*.

Coprinus picaceus has a steeply conical, date-brown cap, which is slightly slimy to the touch and is covered with scattered white felty patches. The long white stalk has no ring. It is an uncommon, but very striking plant, found growing on rich soil, usually in woodlands. Other species of *Coprinus* which grow in woodlands are illustrated on p. 139.

1 Marasmius Oreades   2 Clitocybe Rivulosa   3 Clitocybe Dealbata
4 Conocybe Tenera
5 Coprinus Comatus   6 Coprinus Atramentarius   7 Coprinus Plicatilis

1 **Calvatia gigantea** (Giant Puff-ball, Tête de mort) is found in pastures, sometimes growing in rings, and in woodlands, in late summer and early autumn. It frequently grows to the size of a man's head and sometimes much larger: the record specimen, found in New York State in 1877, was mistaken for a sheep at a distance. It has almost no stalk and is roughly spherical in shape; the surface is downy when very young, but soon becomes smooth like a white kid glove, and finally develops a yellow tinge, somewhat resembling chamois leather. The inside of a young specimen is solid, and the white, spongy flesh is edible; unsuccessful attempts have been made to cultivate the fungus for food in Denmark. With age, the flesh changes to greenish yellow, and eventually forms a mass of olive-brown spores. It has been calculated that an average specimen contains seven million million spores. Eventually the outer skin breaks away at the top, exposing a brittle inner protective layer, which soon collapses and releases the spores. The slightest blow on the fungus then causes the spores to puff out in a cloud; and since they are not readily wettable, rain drops falling amongst them will also cause puffing. Ripe or nearly ripe specimens were at one time used as a country remedy to staunch bleeding, either applied to the wound as a pad, or made to smoulder and then applied to cauterize the wound. They were also used as tinder for kindling fire, and for producing smoke for driving bees from one hive into another.

Calvatia caelatum grows in sandy pastures in summer and autumn, and is the shape and size of a pear. It has a short stalk region, separated inside from the olive-brown spore mass by a distinct skin. The whole outer surface is at first whitish-grey and warted, but soon becomes pale brown and develops a pattern of hexagonal cracks. When ripe, the upper part tears away to form a large opening, and the spores are dispersed by chance blows and rain-splash.

2 **Calvatia excipuliforme** is found in pastures, on heaths, and in woods in late summer and autumn. It has a tall, stout stalk, which is often grooved or furrowed at the top, and the surface of the round, spore-producing part is covered with small spiny warts. The whole plant is white at first, but before the skin of the upper part has peeled away to release the olive-brown spores, it has become pale brown and smooth.

3 **Bovista nigrescens** is found in grassland at all times of the year. Young plants are pale whitish brown, but the outer skin soon peels away to show a greyish-white inner layer. The whole of the inside develops into spores, which are purplish brown when ripe. Then the skin becomes similarly coloured, and eventually tears open to release the spores. Dispersal is aided by the whole plant becoming detached and being rolled about by the wind.

Bovista plumbea, which is much like *B. nigrescens*, is found only in late summer and autumn. The outer layer is whitish grey, but soon cracks and peels off to show the inner skin, which becomes a lead-grey colour with age. The olive-brown spores are dispersed through a rather small tear at the top, and by the plant being blown about by the wind.

4 **Lycoperdon depressum** grows amongst grass in summer and autumn. It is white when young and covered with spiny and granular scales; these gradually fall off as, with age, the colour changes to yellowish brown. Only the upper part produces spores, the lower third of the plant forming a short solid stalk which is separated inside from the spore mass by a distinct skin. When ripe, the spores are olive green, and they are released through a large opening in the top. Other species of *Lycoperdon* are illustrated on p. 155.

5 **Clavulina rugosa** is found in the latter half of the year, growing in meadows and grassy places. It is white, often tinged grey or yellow, and usually has several rather irregular branches, though sometimes it is unbranched. The surface is slightly rough, and the flesh is brittle. A smaller species, *C. cristata*, is illustrated on p. 153.

6 **Clavulinopsis corniculata** grows amongst grass from summer until early winter. It has bright yellow branches and firm, waxy flesh. An orange-yellow unbranched species, *C. helvola*, is illustrated on p. 153.

1 Calvatia Gigantea
2 Calvatia Excipuliforme
3 Bovista Nigrescens
4 Lycoperdon Depressum
5 Clavulina Rugosa
6 Clavulinopsis Corniculata

1 **Brachythecium rutabulum** is a very common moss in grassland and elsewhere, and also on rocks and dead wood. It frequently grows in poorly-drained lawns and is regarded as a pest. It is bright green, sometimes with a yellowish tinge, and has a glossy texture. The creeping, irregularly-branched stems are covered with widely-spreading oval leaves (1A), which have distinctly, although minutely, toothed margins and taper to sharp points. The nerve extends for rather more than half the length of the leaf. Long stalks, rough to the touch, carry the curved spore capsules, which have short, conical lids.

2 **Brachythecium velutinum** is rather like a small form of *B. rutabulum*, and grows in similar places. It forms bright-green glossy patches of irregularly-branched stems with spreading leaves. It can be confused with *Eurhynchium confertum* (p. 182), but the leaves are longer and narrower and have longer, finer points (2A). The margins are minutely toothed, and the nerve runs to just beyond the middle of the leaf. The spore capsules are rather short and have short, conical lids; the stalks are rough to the touch.

3 **Brachythecium albicans** grows amongst grass on sandy soils, and also on bare gravelly ground and on sand dunes. It has numerous, upright, pale yellowish-green branches arising irregularly from a creeping stem and having a silky texture. The leaves, which are broadly oval and contracted abruptly at the tip into a fairly long, fine point (3A), are marked with lengthwise folds and have smooth margins. The nerve extends for rather more than half the length of the leaf. The smooth-stalked, short, oval spore capsules are only rather rarely produced. Other species of *Brachythecium* are shown on pp. 59 and 99.

4 **Rhacomitrium canescens** is found amongst grass in heathy places, and by roadsides and on wall-tops. It forms greyish-green tufts of irregularly-branched stems

with short side branches. The narrow oval leaves (4A) have well-developed nerves running right to the tips and usually taper into long white 'hair points', although sometimes these are poorly developed. The small spore capsules grow on long, smooth stalks. *R. lanuginosum*, a larger, mountain moss, is described on p. 58.

5 **Leptobryum pyriforme** forms short turfs in open places on well-drained soils and sometimes on sandstone rocks. This moss is especially common in 'man made' situations, such as cinder-heaps and on the soil in flower pots. Each plant consists of an unbranched upright stem (5A) with narrowly oval lower leaves tapering into lightly-toothed points, and upper leaves in which the toothed points are three or four times as long as the rest of the leaf. Both kinds of leaf have broad nerves running right to the tips. The pear-shaped spore capsules are held horizontally or slightly drooping on long, orange-red stalks, and become glossy reddish brown when ripe.

6 **Physcomitrium pyriforme** ('Pear Moss') colonizes damp ground and often forms extensive turfs on bare mud patches. The oval leaves have sharp points at the tips and toothed margins. They are small at the base of the plant, but much larger near the tips of the stems. The spore capsules are shaped and coloured like small green pears, and are held erect on short stalks. At first glance the plant can be confused with *Bartramia pomiformis* (p. 60), although the latter does not grow in the same situations.

7 **Lophocolea bidentata** is the commonest liverwort found growing amongst grass, and is frequent on damp lawns, as well as in pastures and grassy places generally. The pale green leaves are set obliquely on the stem in two ranks, and have a pair of prominent teeth at their tips. There are also much smaller, deeply divided underleaves. Other species of *Lophocolea*, which grow on wood, are described on p. 174.

1 Brachythecium Rutabulum   2 Brachythecium Velutinum   3 Brachythecium Albicans
4 Rhacomitrium Canescens   5 Leptobryum Pyriforme
6 Physcomitrium Pyriforme   7 Lophocolea Bidentata

MOSSES AND FUNGI

1 **Funaria hygrometrica** forms extensive pale-green carpets of moss in open places in fields and on heaths, especially on the ashes of fires. Each plant has a short stem and a few broadly oval, pointed leaves, each with a nerve that does not extend quite to the tip. The spore capsules, which are pale orange or yellowish brown when ripe, grow on long stalks and are turned slightly downwards.

2 **Anthracobia macrocystis** is a Cup Fungus (p. 150) which is found commonly on burnt ground in summer and early autumn. The inside of the cup is reddish orange, and the outside is pale orange, often with brown spots near the margin. There are several other similar species, which can be distinguished with certainty only by examining the spores under a microscope.

3 **Morchella esculenta** (Morel) grows in spring on rich soil, especially where wood or paper have been burnt. It has a stout, yellowish-white stalk. Firmly attached to this is a 'cap' criss-crossed with pale-brown ridges separating irregular, shallow hollows lined with the darker-brown, spore-producing surfaces. The plant is edible but should be blanched in boiling water before being prepared for the table. There are several other similar species: *M. elata*, a rare plant of conifer woods in northern Britain, has a smaller, more conical cap with dark grey or black vertical ribs; *Mitrophora semilibera*, quite common in woodlands, has a small conical cap fitting loosely on, but not attached to, a long, narrow stalk; and *Verpa conica*, found fairly frequently in grassland, has a ridged conical cap, with the spore-producing surface on the ridges as well as in the hollows, fitting loosely on a short, stout stalk.

4 **Splachnum ampullaceum** grows in tufts on cattle dung in wet places. This moss is related to *Funaria hygrometrica* and is similar to it in general appearance, but the oval leaves are tapered at the base and at the tip, and the upper parts of the margins are bluntly toothed. The spore capsules, which are held erect on long stalks, have a cylindrical upper part which contains the spores, and a large swollen base, making them look like tiny antique wine-jars.

5 **Panaeolus sphinctrinus** is found in summer and autumn growing on dung. The cap is sooty black when wet, but dries out to a lead grey colour; it is dome-shaped, slightly pointed at the top, and when young, the margin has a fringe of white delicate 'teeth' which drop off with age. The spores are dark brown and under the microscope can be seen to be lemon-shaped. A similar species, *P. foenisecii*, often found on lawns, has a dark brown cap with a reddish tinge and a darker band near the margin, but no 'teeth' on the edge. *P. semiovatus* grows on dung; the whole plant is pale brownish-grey, and there is a well-developed ring on the stalk.

6 **Poronia punctata**, which grows on horse dung, is shaped rather like a nail, with a tapering stalk buried in the dung, and a flat 'head' dotted with the tiny black openings of flask-shaped, spore-producing structures. The botanist R.W.G. Dennis, in his book *British Cup Fungi*, says that the fungus was evidently abundant in the 19th century; but it is now quite rare.

7 **Peziza vesiculosa** is a Cup Fungus (p. 150) which grows on manure heaps and richly manured soil. The pale brown cups have inturned margins, and the inner spore-producing surface is frequently raised up to form blisters in the centre.

8 **Coprinus niveus** grows on cattle dung from late spring until autumn. When young, the cap and stalk are densely covered with small chalky-white scales and have a mealy appearance. With age, the cap becomes split and turned upwards at the margin, and then it and the gills gradually dissolve into a black, inky fluid containing the spores.

9 **Coprinus radiatus** is a small species which is common on fresh horse dung from late spring until autumn. The grey cap is covered with paler, pointed, hairy scales. The white stalk is softly hairy at the base and has a root-like extension below.

10 **Coprinus cinereus** grows on dung at all times of the year. When young, the whole cap is covered with small whitish-brown scales; these fall off with age, except in the centre, where they remain after the rest of the cap and the gills have dissolved away. The stalks are swollen at their bases, where they may be joined together, and have tapering root-like extensions. Other species of *Coprinus* are shown on pp. 35 and 139.

1 Funaria Hygrometrica   2 Anthracobia Macrocystis   3 Morchella Esculenta
4 Splachnum Ampullaceum   5 Panaeolus Sphinctrinus
6 Poronia Punctata   7 Peziza Vesiculosa   8 Coprinus Niveus
9 Coprinus Radiatus   10 Coprinus Cinereus

Grassland communities include comparatively few flowerless plants. Ferns are scarce because grasslands are usually too much exposed to sun and wind, though bracken (p. 185) can invade all kinds of grasslands except those on chalky soils, and *Ophioglossum vulgatum* (p. 191) is sometimes found on chalk downs. Since competition between the plants is intense, in these communities only robust and strongly-growing mosses and liverworts (pp. 38, 42, 44) are able to survive, and lichens (p. 44) are scarce. Fungi, which do not need to compete for light and air, are more numerous (pp. 30-37).

1 **Pseudoscleropodium purum** is a common moss of chalk grassland, also found on heaths and in woodlands, and forms pale-green wefts of robust branches amongst grasses and other flowering plants. The main stem and the short side branches are clothed with broad overlapping leaves, which have rounded tips with short protruding points (1A). This species has been called *Brachythecium purum*, and the fine teeth on the margins of the leaves, the slight folding of their surfaces, and the nerves running about half way to the tip are like those of *B. rutabulum* (p. 39); but the shape is different and the protruding points are distinctive. The plant looks rather like *Pleurozium schreberi* (p. 49), but the stems are pale green and not bright red. It is frequently used by anglers as packing in their bait cans.

2 **Rhytidiadelphus triquetrus** grows in clearings in woodlands on chalky soils, and sometimes on sand dunes, heaths, and moors, as well as on chalk downs. The stiff, reddish stems branch in all directions and form robust bushy wefts. The triangular leaves are widely spreading (2A), and have toothed margins, strongly folded surfaces, and short, though well-defined, double nerves. When dried, the plant looks well in 'floral arrangements'.

3 **Rhytidiadelphus squarrosus** is found in chalk grassland, pastures, and lawns. The reddish, rather irregularly branched stems form extensive wefts, which are less bushy and more slender than those of *R. triquetrus*. The widely-spreading leaves have long, channelled points which are turned backwards (3A). The margins are toothed, and there is a short double nerve; but the surface is only lightly folded. A related species, *R. loreus*, which is found in woodlands in the north and west of Britain, is illustrated on p. 179.

4 **Camptothecium lutescens**, which grows only on chalk and limestone soils, forms yellowish-green loosely-tufted wefts. The slender main stems and the short side branches are clothed with long, narrowly triangular leaves (4A), which have lightly-toothed margins, strongly-folded surfaces, and nerves extending three-quarters of the way or more towards the tip. The plant is rather like *Brachythecium albicans* (p. 39), which takes its place on soils that are free from chalk, but it is rather larger and more robust, and the leaves are longer and narrower and more deeply folded. A related species, *C. sericeum*, which grows on walls, is shown on p. 75.

5 **Hypnum cupressiforme var. lacunosum** is found in chalk grassland. Its rich green colour is tinged with yellow as shown here, or may be a striking golden green, sometimes with deep bronze tints. It is the most robust form of a very variable species and, unlike the typical variety (p. 177), the shoots are not at all flattened but have stout, upright stems with short, rather irregularly arranged branches. The crowded, overlapping, curved leaves have short points and are turned to one side (5A); their boat-like shape gives the shoots a swollen appearance.

Hypnum cupressiforme var. tectorum, which grows in grassland on all types of soil and also on walls and roofs, is similar to var. *lacunosum*, but not quite so large. It is dark or olive green, and the leaves are only very slightly curved.

6 **Acrocladium cuspidatum** is quite common in chalk grassland, and also grows in an entirely different kind of habitat, in very wet places on boggy ground. The bright yellow-green shoots have pointed, spear-shaped tips, due to the young leaves being tightly rolled together. The leaves are oval, tapering from a wide base to a bluntly rounded tip, and they usually have a very short double nerve, though this may be entirely missing. Those on the short side branches are smaller and narrower than those on the main stems. A related species, *A. sarmentosum*, is shown on p. 91.

1 Pseudoscleropodium Purum   2 Rhytidiadelphus Triquetrus
3 Rhytidiadelphus Squarrosus   4 Camptothecium Lutescens
5 Hypnum Cupressiforme *var*. Lacunosum   6 Acrocladium Cuspidatum

1   **Cladonia pocillum** is a lichen which forms continuous crusts of small 'leaves' or squamules on the surface of the ground or over mosses in grassland on chalky soils. Each squamule (1A) is brownish green on the upper surface, and white below. From them grow grey-green, stalked cups with warted surfaces. This species was at one time regarded as a variety of *C. pyxidata* (p. 28).

2   **Cladonia subrangiformis** grows amongst grasses on chalk and limestone soils. The greenish-grey, slender, hollow stems are repeatedly and rather irregularly forked, and there are always some 'scales' or squamules at the base, or scattered elsewhere on the plant. It is similar to *C. furcata* (p. 51), but the branches are set at much wider angles, and there are round white spots on the main stems near the base.

3   **Ditrichum flexicaule** is a moss forming dense, dark-green tufts on chalk downs and limestone hill grassland. The stems are very thin and thread-like, and the pointed leaves are curved and turned to one side. Each leaf has a broad base, immediately above which it narrows suddenly and then tapers gradually to a long fine point, which has minute teeth near the tip. There is a broad nerve running the whole length of the leaf, and a narrow border at the margins.

4   **Ctenidium molluscum** is a moss which grows in extensive wefts or mats of very neatly and regularly branched 'fronds'. The leaves on the main stems are broadly heart-shaped with long, strongly-curved tips, while those on the branches are smaller and narrower, but equally curved (4A). They have finely-toothed margins but no nerve. On chalk downs the plant is golden green in colour, but another form, to be found on wetter, less chalky soils, is bright green without any yellow tinge.

5   **Neckera crispa** is a moss found in open chalk grassland, on limestone rocks, and sometimes, in very wet places, on the bases of trees. It has branched, very much flattened shoots, which have a glossy texture and are frequently curved and curled at the tips. The leaves have a rounded oblong shape with a tiny point at the tip, and are markedly transversely folded (5A). Usually there is a very short double nerve, but this may be missing altogether.

6   **Anomodon viticulosus** grows in shaded places on chalky soils and on limestone rocks and walls. This moss has long creeping stems with short upright branches. When the plant is moist, the leaves are bright green and spread widely (6A), but when dried they become much twisted and pressed against the stem and change to a very dull green. Each leaf is broadly oval for about two thirds of its length and then narrows to a short strap-shape with a blunt tip. The nerve extends nearly the whole length of the leaf.

7   **Thuidium philiberti** forms loose wefts of moss in chalk grassland. The long creeping stems bear short side branches, which branch again, and are clothed with small heart-shaped leaves with long tips (7A). The nerve extends almost to the tip of the leaf. The plant is very similar to *T. tamariscinum* (p. 179), but the latter is usually more luxuriously branched, and has leaves on the main stems which do not have the long tips.

Thuidium abietinum is common in grassland on chalky soils. The main stems branch only once, but the strongly nerved, heart-shaped leaves with elongated tips are like those of other species of *Thuidium*.

8   **Eurhynchium swartzii** forms yellowish-green wefts or mats in open chalk grassland, and has a glossy appearance when dry. The long main stems are rather irregularly branched, and clothed in broad, heart-shaped leaves (8A). The nerve extends to just beyond the middle of the leaf, and the margins are sharply toothed. This moss is very similar to *E. praelongum* (p. 183), but the latter is more regularly branched, and the leaves on the branches are markedly smaller than those on the main stems.

9   **Eurhynchium striatum** grows in loose wefts on chalk soils, and is often, though not always, found in shaded places in woods and on hedge banks, as well as amongst grass. It has long, irregularly-branched stems, and triangular leaves which are strongly folded along the length when dry. The nerve extends for about three quarters of the length of the leaf, and the margins are sharply toothed. As in other species of *Eurhynchium* the spore capsule is curved and has a lid with a conspicuous curved beak.

1 Cladonia Pocillum   2 Cladonia Subrangiformis   3 Ditrichum Flexicaule
4 Ctenidium Molluscum   5 Neckera Crispa   6 Anomodon Viticulosus
7 Thuidium Philiberti   8 Eurhynchium Swartzii   9 Eurhynchium Striatum

HEATHLAND FUNGI

Heath vegetation develops on poor soils where trees will not grow because of altitude, exposure to strong winds, or grazing by animals. The dominant flowering plant is usually heather, and bracken is often abundant. Heaths are normally better drained than moors, and peat, if it is formed at all, only accumulates in thin layers. Although heaths are often in upland areas, they are also to be found in the lowlands of southern and eastern England. Most of the plants described here grow on moors as well.

1 **Marasmius androsaceus** ('Horsehair Toadstool') grows on small sticks, especially old heather stems, and on conifer needles, in late summer and autumn. The finely wrinkled cap is pale reddish brown and has a hollow in the centre which is darker in colour; the gills are rather distantly spaced. The stalk is long, thin, stiff, and shiny black like a horse hair. Another species, *M. rotula*, growing on sticks in woodlands, has a similar stiff, shiny black stalk, but a whitish-grey, grooved cap. *M. ramealis*, also found on twigs, is shown on p. 145.

2 **Galerina hypnorum** is found amongst mosses in damp places on heaths, in woodlands, and also in bogs, in summer and autumn. The cap is pale yellow when wet, but turns darker on drying; it is cone shaped, and the margin is grooved almost to the centre. The gills are broad and distantly spaced, and the delicate stalk is tinged rusty red towards the base.

3 **Hebeloma crustuliniforme** ('Fairy Cakes') is a widespread species, appearing on damp soil on heaths and elsewhere in late summer and autumn. The whole plant is a pale whitish-brown colour. The margin of the cap is inrolled and covered with very short soft hairs. The gills are watery, and the cap is slightly slimy to the touch. The short, thick stalk is covered with a coarse white powder at the top. The firm white flesh, with a distinct smell of radishes, should not be eaten as it has been known to cause illness.

4 **Hebeloma mesophaeum** grows on heaths and also in open woodlands, especially associated with pine and birch, from late summer until early winter. The cap is dark brown in the centre but much paler at the margins, and the gills are covered by a veil at first, which soon breaks away, leaving a distinct ring on the stalk. Both gills and stalk are white when young, but become brown with age.

5 **Coltricia perennis** is found throughout the year in open places on light soils, especially where the ground has been burnt. The cap is funnel shaped and rusty brown in colour, with alternating lighter and darker zones, and it has a covering of fine silky hairs when young. The brown stalk is covered with very short hairs. The yellowish-brown spores are produced on the underside of the cap in tubes which have very small openings and are yellowish grey when young, but become dark brown with age. The flesh is brown and very tough and leathery in texture. The plant is related to *Polyporus* (p. 128) rather than to *Boletus* (p. 142), because the tubes are not easily separable from the flesh of the cap.

6 **Omphalina pyxidata** grows amongst grass on moist, light soils in late summer and autumn. The brown cap is usually tinged with orange, and sometimes streaked with reddish tints, and is hollowed in the centre, with a grooved margin. The stalk is paler coloured and wavy, and the gills, which run down the stalk, are yellowish brown. Other species of *Omphalina* are shown on pp. 85 and 89.

7 **Pholiota carbonaria** ('Charcoal Toadstool') is associated with conifers and found in autumn, especially on burnt ground. The whole plant is yellowish brown. The cap has a thin skin which can fairly easily be peeled off, and is slimy to the touch. There are scattered, cottony hairs on the lower part of the stem.

8 **Thelophora terrestris** is found from late summer until early winter in open places on heaths and also in conifer woods. It forms dark-brown, irregular, fan-shaped masses with paler, fringed margins. The lower, spore-producing, surface is dark grey and wrinkled. The whole plant may be flushed with violet tints.

1 MARASMIUS ANDROSACEUS   2 GALERINA HYPNORUM   3 HEBELOMA CRUSTULINIFORME
4 HEBELOMA MESOPHAEUM   5 COLTRICIA PERENNIS   6 OMPHALINA PYXIDATA
7 PHOLIOTA CARBONARIA   8 THELOPHORA TERRESTRIS

1 **Ceratodon purpureus** is a moss forming extensive turfs on bare soil on heaths and moors, often amongst heather. It also colonizes burnt ground and other artificial situations such as fallow fields and wall tops. Spore capsules (1A), with curved conical lids and conspicuous purplish red stalks, are produced abundantly in the spring. The capsules grow upright at first, but when ripe they become dark reddish brown and deeply grooved and are held almost horizontally. The erect stems bear spreading, narrow, triangular leaves, which taper to fine points and have inrolled margins. Each leaf has a well-developed nerve running in a channel for its entire length.

2 **Pohlia nutans** is a moss growing on peat, on decaying wood, and on sandy ground, but never on chalk or limestone soils; it forms dull, deep-green turfs. The spore capsules, which are a very bright green when young, have long red stalks and hang downwards. The leaves on the upper part of the stem (2A) taper to a faintly toothed point, but the lower leaves are shorter and more oval. The nerve does not extend quite to the tip.

Pohlia delicatula is usually found on clay soils in moist situations, and has pale green leaves which are bluntly but more distinctly toothed than those of *P. nutans*, and with nerves that extend for only three-quarters of the length of the leaf. The stems and leaf bases are red, and the spore capsules hang downwards and have short, rather fleshy stalks.

3 **Pleurozium schreberi** commonly grows amongst heather on heaths and mountain slopes, amongst pine trees, and in upland oak woods. This moss is never found on chalk or limestone soils. It forms large, light-green, tangled wefts of stiff upright main stems, with many somewhat curved and rather irregularly arranged branches. The broadly oval leaves have very short, inconspicuous, double nerves and inrolled margins which give them a hooded appearance; those on the main stems are twice as large as those on the branches (3A). At a first glance this plant is rather like *Pseudoscleropodium purum* (p. 43), but it may be clearly distinguished by its bright red stems. It also somewhat resembles *Acrocladium cuspidatum* (p. 43), but the latter has much more pointed tips to its branches.

4 **Hypnum cupressiforme var. ericitorum** grows on heaths and in upland woods, and is usually associated with heather. It is more slender than the typical variety of this very variable moss (p. 177), and forms pale whitish-green wefts of upwardly-growing stems rather than flattened mats. Also the leaves are less crowded on the very regularly branched stems, but they are strongly curved to one side, which give the tips of the branches a hooked appearance.

5 **Lophozia ventricosa** is a leafy liverwort which forms bright yellowish-green patches on sandy and peaty soils, and on rotten wood. It is commonest in the north and west of Britain. The brown spore capsules emerge from conspicuous sheaths or perianths at the base of the stalks (5A). The overlapping leaves are shallowly notched at their tips, and the uppermost ones have clusters of pale green, powder-like reproductive bodies (gemmae) on the points.

6 **Diplophyllum albicans** is a common liverwort on all types of soils except chalk and limestone, and it often covers quite extensive areas of ground. The turfs of crowded, upright shoots growing from creeping stems vary from pale green, often with a whitish tinge, to reddish brown. The leaves are closely set in two ranks, and each consist of two oblong lobes which are faintly toothed at the tip (6A). There is a narrow white band running down the middle of each lobe, which is rather like the vein in the leaf of a moss. Reddish-brown spore capsules are commonly produced, and also clusters of yellowish-green, powdery reproductive bodies (gemmae) on the margins of some of the upper leaves.

7 **Lycoperdon ericitorum** grows in late summer and autumn on heathy soils. It is usually slightly pear-shaped, and is greyish white and the upper part pale yellowish brown. The surface is covered with very small scales. The whole of the inside of the fungus develops into a mass of cinnamon-coloured spores, which escape through a small raised pore at the top. Other species of *Lycoperdon* are described on p. 154.

1 CERATODON PURPUREUS   2 POHLIA NUTANS   3 PLEUROZIUM SCHREBERI
4 HYPNUM CUPRESSIFORME *var.* ERICITORUM   5 LOPHOZIA VENTRICOSA
6 DIPLOPHYLLUM ALBICANS   7 LYCOPERDON ERICITORUM

1 **Cladonia arbuscula** forms dense, bushy clumps on heaths and moors and in the drier parts of bogs throughout Britain. The hollow stems are much branched, the branches often developing in groups of four; and the tips, which bear small, brown, spore-producing structures, are turned over in one direction, giving a comb-like appearance (1A). The plant never develops any 'scales' or squamules on the stems, such as are always found, for instance, on *C. furcata*. It is always tinged yellow, and has a slightly bitter taste, due to the presence of usnic acid, an antibiotic drug.

Cladonia rangiferina (Reindeer Moss) is an uncommon plant in Britain, found mainly at fairly high altitudes in Scotland and in parts of Wales. It is very like *C. arbuscula* in shape but is usually taller and more sparingly branched in its lower parts, bluish grey without any yellow tinge, and without the bitter taste. If the stems are touched with a drop of strong potassium hydroxide solution, a lemon-yellow colour will develop. In arctic and subarctic regions, where this lichen is abundant, it forms the staple diet of reindeer.

2 **Cladonia furcata** grows on the ground on heaths and in open woodlands, and sometimes on rotting tree stumps. The brownish or greenish grey, slender, hollow stems are repeatedly forked, and there are always some 'scales' or squamules at the base, or scattered elsewhere on the plant. Most of the branches end in sharp points which are tinged brown, but some have groups of small, dark-brown, spore-producing apothecia (2A). There is an opening into the hollow of the stem at the base of each branch. A related species, *C. subrangiformis*, which grows in chalk grassland, is shown on p. 45.

3 **Cladonia rangiformis** is found on heaths on all kinds of well-drained soils, and by the sea on sand dunes and on shingle. It forms stiff, tangled bushy clumps of whitish-grey, hollow stems, which are forked many times. There are openings at the bases of the branches, and small 'scales' or squamules are always present, especially towards the bottom of the plant.

4 **Cladonia impexa** is common on heaths and moors, growing in dense cushions. The hollow stems are much branched, the branches often developing in groups of three; but the tips are not turned in one direction, as they are, for example, in *C. arbuscula*. Sometimes small, brown, spore-producing apothecia develop at the tips of the branches (4A). The plant, like *C. arbuscula*, never develops any 'scales' or squamules, and it contains usnic acid and so is always tinged yellow and has a slightly bitter taste.

Cladonia tenuis, a quite common species with a slender and neat appearance, is distinguished from *C. arbuscula* because its branches develop in groups of two, from *C. impexa* because the tips of the stems turn over in one direction, and from *C. furcata* because it never bears any squamules.

5 **Cladonia crispata** usually grows on peat, and forms a layer of scale-like brownish-green 'leaves' or squamules, from which grow rather short upright stems that branch once or twice. At the tips are very small and rather irregular cups which may have small, brown, spore-producing apothecia on their margins. At the bottoms of the cups and where the stems branch there are conspicuous openings into the hollow stems.

6 **Cladonia gracilis** is a tall, slender, handsome plant of peaty soil, with sparingly branched, smooth, and rather shiny stems. The branches either expand into narrow shallow cups, or taper to a point. The cups, which have no openings inside, may give rise to further branches or small, brown, spore-producing apothecia on the toothed margins. The scale-like 'leaves' or squamules at the base are small and not very numerous.

7 **Cladonia uncialis** is variable in size; in wet situations it may be two or three times as large as shown here. The stems are broad and hollow and often have an inflated appearance; they are forked once or twice, and the branches are short and pointed, with brown tips. There are conspicuous openings where the stems branch (7A), and the plant never develops any 'scales' or squamules.

8 **Cornicularia aculeata** is found on dry soils in open places, growing in tangled cushions. The flattened stems are intricately branched, dark reddish brown, shiny, and stiff.

Cornicularia muricata grows in similar places, but is smaller and more slender than *C. aculeata*. The stems are narrower and less flattened and grow upright; and there are small spiny bristles on the upper branches.

1 CLADONIA ARBUSCULA   2 CLADONIA FURCATA   3 CLADONIA RANGIFORMIS
4 CLADONIA IMPEXA   5 CLADONIA CRISPATA   6 CLADONIA GRACILIS
7 CLADONIA UNCIALIS   8 CORNICULARIA ACULEATA

1 **Baeomyces rufus** forms grey-green crusts on peaty, sandy, and gravelly soil, and on damp rocks, especially sandstone. The spore-producing apothecia (shown enlarged in 1A) are reddish brown and dome-shaped, grow on short stalks, and look like tiny toadstools. The stalks, unlike those of *Pycnothelia papillaria* (No. 5) and *Cladonia* species (p. 50) are solid, with no hollow space inside. The typical form of *B. rufus*, as shown here, is a granular crust with an irregular margin and does not form any 'scales' or squamules; but in the quite common variety *subsquamulosus* tiny squamules develop, especially towards the edges.

2 **Baeomyces roseus** consists of a whitish-grey crust, sometimes sprinkled with tiny pink granules, and growing on soil. The spore-producing apothecia (2A) are pink and grow on short stalks. This plant frequently appears on freshly exposed but firm surfaces on banks and the sides of ditches, especially on gravelly soils, and plays a part in consolidating them against erosion.

3 **Lecidea granulosa** is found in damp places on peat, and also on rotten wood, forming a thin layer of light-grey granules. The surface is sometimes lightly sprinkled with a fine yellowish-green powder, which consists of minute reproductive bodies (soredia). The spore-producing apothecia are dark green with a velvety appearance when young, but later change to brick red or dark brown, and eventually become almost black (3A). The 18th-century naturalist Dickson called the plant '*Lichen quadricolor*' because of this.

4 **Icmadophila ericitorum** grows on decaying turf and rotting wood on heaths and moors, and is especially common in Scotland. It forms thick, rather soft crusts which are greyish white when dry, but have a distinctly greenish tinge when wet. The pink spore-producing apothecia are usually slightly irregular in shape and unstalked (4A), which helps to distinguish this plant from *Baeomyces roseus* (No. 2).

5 **Pycnothelia papillaria** is found on peaty heaths, especially in the north and west of Britain. It has a whitish-grey, thin, granular crust, from which grow short, hollow stalks (5A) with small brown, spore-producing structures at their tips. This species is often called *Cladonia papillaria*.

6 **Lecanora gibbosa** grows on stones and is common on flints, as shown here. The dark-grey surface is shiny and sometimes tinged with green, and has a warted appearance. The thinner, almost black margin frequently extends some way as a feathery stain on the rock. The spore-producing apothecia develop within the warts and break through on the surface when fully developed, giving the appearance of minute volcanic craters (6A).

7 **Rhizocarpon obscuratum** forms thin, regular patches on rocks and stones. Each patch is brownish grey, very finely cracked, and surrounded by a dark margin which appears as a stain on the stone. The spore-producing apothecia are black and very small (7A). The typical form is commonest in damp places in the west of Britain, but the variety *reductum*, shown here, is most frequent in the south east in rather drier situations, and especially on smooth flints. It has a very well-developed border.

8 **Hypogymnia physodes** form **elegans** is found on moors and heaths growing on twigs, and is especially common on old heather stems. The branching lobes are narrow and elegant, and turned up at the tips, giving the plants a dainty and almost lace-like appearance. The typical form of *H. physodes* is shown on p. 165.

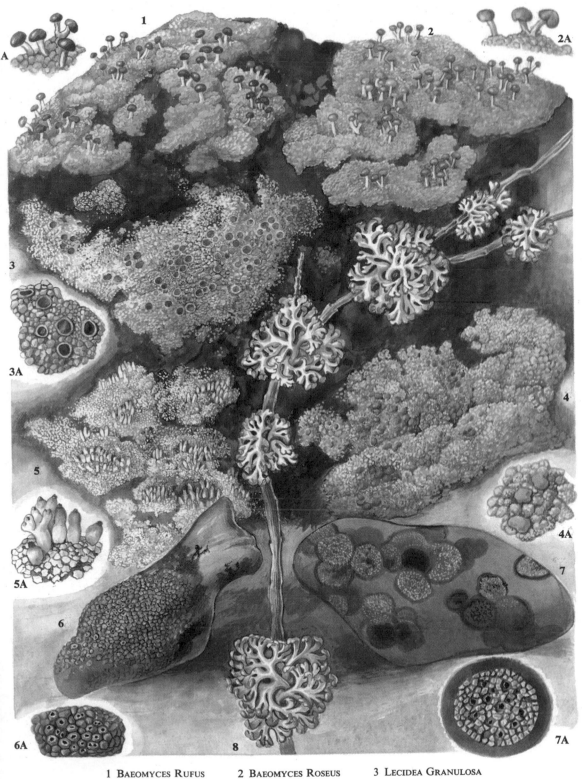

1 Baeomyces Rufus   2 Baeomyces Roseus   3 Lecidea Granulosa

4 Icmadophila Ericitorum   5 Pycnothelia Papillaria

6 Lecanora Gibbosa   7 Rhizocarpon Obscuratum   8 Hypogymnia Physodes *form* Elegans

Although the mountains of Scotland, Wales, and Ireland are only hills compared with, say, the Alps, their vegetation does include a considerable proportion of 'arctic-alpine' species — that is plants which are able to withstand the effects of both bright sunshine combined with low temperatures and steep slopes with severe exposure. They are able to grow only above the limit of tree growth, and the clearing of woodlands from most of the hillsides of Britain has helped them to extend their range to lower altitudes. As well as those described on the following pages of mountain plants, many of the heath and moorland species are also found on mountainsides.

1 **Gymnocarpium dryopteris** (Oak Fern) grows in clefts between rocks, usually on north-facing screes, and also in rocky places in woods and by streamsides. It is quite common in the north and west of Britain, although rare in Ireland, but is never found on limestone. The long-stalked fronds grow singly at irregular intervals from slender, creeping stems which have a covering of small pale-brown, oval scales on their younger parts. The older parts, not shown in the illustration, are black and glossy. The slender and brittle stalks have a few scales at their bases and often grow to nearly three times the length of the rest of the frond. The blade of each frond has from six to twelve opposite pairs of main lobes, which are twice further subdivided. The lowest pair of lobes is much larger and has much more stalk than the others, giving the frond the appearance of being divided into three equal diverging parts. This is strikingly seen when the fronds are very young, for the parts of the blade are tightly rolled up and look like three little balls on wires. Although in the illustration the frond is shown somewhat flattened out, another striking feature is the way in which the whole blade is bent back sharply at right angles at the top of the stalk. The spore cases develop beneath the fronds in circular clusters which are not protected by a scale (1A).

2 **Gymnocarpium robertianum** ('Limestone Polypody') is found on limestone rocks, chiefly in north and west England and in Wales. It is also known in one locality in Sutherland in the extreme north of Scotland, one in Sussex, and one in western Ireland. The creeping stems are dark brown and not glossy, and the younger parts are clothed with pointed brown scales. The fronds, when young, have soft stalks and are covered with small glandular hairs. Later, the stalks become firm, but they do not grow to more than twice the length of the blades. The blade of each frond has from six to twelve opposite pairs of main lobes, which are twice further subdivided. Each pair of lobes is smaller than the pair below it, but the lowest ones are not as large as the whole of the rest of the frond, as they are in *G. dryopteris*, and so the blade does not appear to consist of three equal parts. Nor is the blade bent back sharply at the top of the stalk, though in the illustration the stalk has been broken in order to include the whole frond on the page. The spore cases develop beneath the fronds in circular clusters which are not

protected by a scale (2A). The name *robertianum* refers to the smell of the crushed fronds, which is said to resemble that of Herb Robert (*Geranium robertianum*).

3 **Cryptogramma crispa** (Parsley Fern) colonizes rock-slides and screes, and is found also on dry-stone walls and the spoil-heaps of quarries, although it never grows on limestone. It is very common in the Lake District and in north Wales, but less so in Scotland. It grows in Ireland in the mountains of Mourne, and in one locality on Exmoor. The fronds grow in dense clustered tufts from the growing tip of a creeping stem which is covered with pale-brown oval scales. Those on the outside of the cluster have, as shown in the upper plant in the illustration, pale green, brittle stalks, and bright green blades which are wedge-shaped and divided four times into small lobes with bluntly-toothed margins. The fronds in the middle of the tuft, shown in the lower plant in the illustration (3A), from which the outer fronds have been cut away, are divided only three times, and the lobes are not toothed. Spore cases develop on the undersurfaces of the lobes of these special fertile fronds and are protected by the infolded margins which are turned back over them.

4 **Polystichum lonchitis** (Holly Fern) is chiefly found in the Scottish Highlands on limestone rocks. It also grows rather sparingly in northern England, north Wales, and western Ireland. The stiff, dark-green fronds grow in upright tufts from a short underground stem, which is covered with the persistent bases of old fronds in its older parts and with numerous broad pale-brown scales towards the growing tip. Similar scales grow on the short stalks of the fronds and extend about half way up the blades, which are divided into opposite pairs of toothed leaflets. Spore cases develop in groups (sori), each protected by a delicate, greyish-brown, circular scale which shrivels up with age (4A). They are to be found on the undersides of the leaflets of the upper half of the frond only. The name Holly Fern alludes to the sharply spined teeth on the margins of the tough, shiny fronds, which remain green through the winter. Other species of *Polystichum* are shown on p. 187.

1 GYMNOCARPIUM DRYOPTERIS   2 GYMNOCARPIUM ROBERTIANUM
3 CRYPTOGRAMMA CRISPA   4 POLYSTICHUM LONCHITIS

The Clubmosses belong to an ancient group of plants which, in the age of the coal forests 250 million years ago, included giant trees. Now the only survivors are the Clubmosses shown here and *Isoetes* (p. 96). Although sometimes referred to as 'Fern Allies', they are not closely related to ferns. Two differences are their small, undivided leaves and the spore cases, which grow on the upper sides of the leaf bases. They are also not related to mosses, for they have well-developed, woody tissues, which are never found in mosses.

1 **Botrychium lunaria** (Moonwort) is a fern which grows on mountain ledges and in grassland on hillsides in the north and west of Britain. It used to be found less commonly in the lowlands as well, but has now become very rare in southern England, due to the ploughing up of old pastures. A single frond grows each year from a short, fleshy rootstock. It divides to form a leafy part with from three to eight opposite pairs of fan-shaped leaflets, and a separate, erect, much branched stalk on which grow the spherical spore cases (1A). A related plant, *Ophioglossum vulgatum*, is described on p. 190.

in pairs, but sometimes singly, on slender, upright stems with only a sparse covering of very small leaves. Each spore case is kidney shaped and is enclosed by a sharply-pointed leaf with a very thin and papery texture (3A). The spores form a very fine, bright-yellow powder called 'lycopodium powder', which was formerly used as a constituent of 'flash powder', in fireworks, and by pharmacists for coating pills. A somewhat similar species, *L. inundatum*, is described on p. 92.

2 **Lycopodium selago** ('Fir Clubmoss') grows on mountains in Scotland, northern England, and Wales, and is widespread in Ireland; but it is very rare in lowland England, although it was found recently in Sussex by a streamside in a wood. It usually grows in open, damp places, in clefts between boulders, and on rock ledges. Each plant has a tuft of stiff, erect branches, which are forked once or twice and are covered with sharply-pointed, narrow leaves. Long, wiry roots grow from the bases of the stems. The round spore cases (2A) are scattered rather sparingly amongst the leaves, growing on the upper surfaces of the leaf bases. The spores form a very fine yellow powder. In Britain the plant reproduces chiefly by means of small buds (bulbils), also to be found scattered amongst the leaves. These fall off and are able to grow into new plants.

3 **Lycopodium clavatum** (Stag's-Horn Moss) is found chiefly in Scotland and Wales, but grows also in a few scattered places on heaths in lowland England and Ireland. It has a long, creeping, repeatedly branched stem, densely clothed with leaves, and anchored to the ground by tough, wiry roots which grow at intervals along its length. The leaves have very finely-toothed margins, and long, whitish-green, hair-like tips. The spore cases are produced in cones which usually grow

4 **Lycopodium alpinum** grows only in the highlands of Scotland and on mountains in northern England, Wales, and a few places in Ireland. The long, creeping, much-branched stem is square in cross-section; it is firmly anchored to the ground by numerous wiry roots, and only sparsely covered with small, bluntly pointed leaves. Tufts of upright branches with overlapping bluish-green leaves in four ranks grow at intervals from the creeping stem. Cones develop at the tips of some of them. The round spore cases are enclosed in oval leaves with jagged margins and pointed tips (4A); they produce a fine powder of bright yellow spores.

5 **Selaginella selaginoides** grows in Scotland, northern England, north Wales, and north-west Ireland. This clubmoss has creeping stems with numerous upright branches clothed in small pointed leaves. The roots are few and very fine and delicate. Cones grow at the tips of the branches and have scale leaves, which are similar to the ordinary leaves but about twice as large. *Selaginella* differs from *Lycopodium* in having two kinds of spore cases (5A). The more numerous ones found in the lower parts of the cones each produce four large, pale-yellow spores which form conspicuous bulges in the walls of the ripening cases. Those growing at the tips of the cones are kidney-shaped and produce considerable quantities of very small whitish-yellow spores.

1 Botrychium Lunaria

2 Lycopodium Selago 3 Lycopodium Clavatum 4 Lycopodium Alpinum
5 Selaginella Selaginoides

1 **Rhacomitrium lanuginosum** is often the dominant plant over considerable areas on mountain tops, forming a characteristic vegetation known as 'Rhacomitrium heath'. It is rather slow growing, but is able in time to cover rock and peat surfaces with very extensive, ragged, grey-green mats. The long stems branch at intervals, and also bear many shorter secondary branches. The leaves are long and narrow and have white 'hair points', often longer than the rest of the leaf, which have strongly-toothed edges. There is a well-developed nerve which runs into the hair point. Spore capsules, which are not very frequently produced, are small and egg-shaped, with short rough stalks. *R. canescens*, a lowland species, is described on p. 38.

2 **Scapania undulata,** the only plant not a moss on this page, is a liverwort common in the north and west of Britain, where it grows on mountain ledges, on wet ground by springs, and in swiftly-running streams. It may also be found in and by woodland streams in the south east. Some forms of the plant are reddish purple and have finely-toothed leaves, while others have no teeth and are a bright vivid green. The leaves are folded, the smaller lobes being slightly more than half the size of the larger ones (2A). Sometimes minute, pale-green, powdery reproductive structures (gemmae) develop at the tips of the upper leaves. A related species, *S. nemorosa*, is described on p. 182.

3 **Hyocomium flagellare** grows on mountains on the banks of streams and by waterfalls. It forms loose, bright-green tufts, with a golden tinge at least at the tips, hanging down the surface of the rock. The long stems are usually sparingly branched, but some forms have regularly-arranged side branches in opposite pairs. The leaves are heart shaped and toothed at the edges, and there is no nerve (3A).

4 **Bryum pseudotriquetrum** is found on wet ground by streams on mountainsides, and less commonly in marshy places in the lowlands. It forms dense green tufts which are usually tinged with reddish purple. The lower parts of the stems are matted together with a covering of short brown hairs. The oval and pointed leaves have well-developed nerves running into the tips

and margins with a thickened border. Other species of *Bryum* are described on pp. 75 and 83.

5 **Anomobryum filiforme** grows in rock clefts and on gravelly soil in mountainous districts. The unbranched stems are clothed with incurved, overlapping leaves, which give the whole shoot a smooth outline (5A). Each leaf is small, broadly oval, and with a nerve not quite reaching the tip.

6 **Philonotis fontana** is found in boggy ground surrounding mountain springs. It is bright yellowish green, and has long sparingly branched stems which are matted together in their lower parts with a covering of short, reddish-brown hairs. The small triangular leaves have a well-developed nerve and several lengthwise folds (6A). The round, furrowed spore capsules grow on long, orange stalks. Male reproductive structures develop at the ends of the shoots on separate plants in almost flower-like 'buds' consisting of leaves with rounded tips.

7 **Dicranella squarrosa** forms brilliant golden-green tufts or mats in wet situations on mountainsides. The leaves are bent back from the stems (7A) and have nerves not quite running the whole length of the leaf. The rather rarely-produced spore capsules are held almost upright on short, stout, bright red stalks. A related species, *D. heteromalla*, is described on p. 182.

8 **Brachythecium plumulosum** grows on wet rock ledges on mountains and on boulders and occasionally on wood in lakes and streams, mostly in upland districts. It forms wide, dark-green mats, with a golden-brown tinge and a silky texture. The oval leaves are obliquely set on the irregularly-branched stems, and are turned to one side at the tips of the branches (8A). They overlap one another, and the toothed margins are slightly incurved. The nerve extends for just over half the length of the leaf. The curved spore capsule grows on a long stalk, which is smooth for most of its length, but rough to the touch at the top. Other species of *Brachythecium* are described on p. 39.

1 Rhacomitrium Lanuginosum

2 Scapania Undulata   3 Hyocomium Flagellare   4 Bryum Pseudotriquetrum

5 Anomobryum Filiforme

6 Philonotis Fontana   7 Dicranella Squarrosa   8 Brachythecium Plumulosum

59

1 **Hymenophyllum tunbridgense** ('Filmy Fern') is widespread in the western parts of Britain, as far north as the south-eastern part of the Isle of Skye. It is also found in Sussex and, as its name implies, it used to be found in Kent, but the activities of collectors during the past hundred years have eradicated it there and greatly reduced its numbers elsewhere. It grows on rocks and also on tree trunks, but it needs a very moist atmosphere. The delicate fronds, which grow from tough, wiry, creeping stems, are only one cell thick and are repeatedly forked; the veins in the lobes do not extend quite to the tip. The spore cases grow in clusters (sori) which are enclosed in a pair of scales with toothed margins (1A).

Hymenophyllum wilsonii is also widespread in western Britain and extends to the extreme north of Scotland, but it has never been found in the south east. The fronds are rather one-sidedly branched, and the tips and margins of the lobes are bent back. The veins extend right to the tips of the lobes. The margins of the scales enclosing the spore cases are not toothed.

2 **Tortella tortuosa** commonly forms dense, rather pale-green cushions in hollows on limestone rocks in upland areas, and may also be found in hillside grassland on limestone soils. It has long, narrow leaves tapering gradually to fine points, with nerves that run almost to the tip, and wavy margins (2A). They become greatly curled and twisted when dry.

3 **Andreaea rupestris** colonizes hard rock surfaces on mountains, forming dark reddish-brown or almost black patches. The spore capsules (3A), which open by four slits instead of by a lid, are unlike those of any other moss. *A. alpina* is similar, but two to three times as large. *A. rothii* differs from the other two species in that the leaves have well-developed nerves. None of the three grows on limestone.

4 **Hedwigia ciliata** forms pale greyish-green tufts on rocks, other than limestone, in hilly districts. The long stems are repeatedly forked and also bear short side branches. The leaves (4A) have a conspicuous toothed 'hair point', and there is no nerve.

5 **Cratoneuron commutatum** grows by streams and waterfalls on limestone rock in mountain districts, and sometimes in fens in the lowlands. The long stems have very regularly arranged, short, side branches, and the leaves are curved to one side (5A), so that the plants look like golden-green miniature ostrich plumes. The leaves have a well-developed nerve which does not extend quite to the tip, and there are branched thread-like structures (paraphyllia) interspersed amongst them. Another form of *C. commutatum*, which grows in boggy places on high moorland, has larger leaves and is less regularly branched.

6 **Bartramia pomiformis** ('Apple Moss') forms pale bluish-green tufts on rock ledges on mountains, and also on sandy banks in lowland districts. The lower parts of the stems are covered with short, reddish-brown hairs, and the narrow leaves have long, finely-toothed points. The well-developed nerve does not quite extend to the tip of the leaf. When not quite ripe, the spore capsules are shaped and coloured like tiny green apples. At a first glance this plant can be confused with *Physcomitrium pyriforme* (p. 39), although the latter does not grow in the same situations.

7 **Amphidium mougeotii** grows in large, dark-green, compact and rounded cushions on wet rock faces on mountains. The individual plants (7A) have long, narrow leaves with well-developed nerves not extending quite to the tips. The leaves are spread out at a wide angle when wet, but become spirally twisted when dry.

8 **Schistostega pennata** is found in the entrances to caves and old mine shafts and in crevices in sandstone rock, especially in the West Country. It can grow only where the light intensity is low. The delicate, broad, flat leaves grow vertically in two rows from the stems, as do the leaves of species of *Fissidens* (p. 181); but in the latter there is a well-developed nerve, and the base of the leaf has a characteristically folded shape. The stems of *S. pennata* grow from a weft of fine threads (protonema). These have a remarkable power of reflecting light and so appear to glow with a golden-green lustre in a very striking manner. The spore capsules (8A) are very small and grow from shoots which have only a few leaves clustered at the tip.

1 Hymenophyllum Tunbridgense   2 Tortella Tortuosa   3 Andreaea Rupestris

4 Hedwigia Ciliata   5 Cratoneuron Commutatum

6 Bartramia Pomiformis   7 Amphidium Mougeotii   8 Schistostega Pennata

1 **Parmelia omphalodes** grows rather loosely attached to rocks in mountainous districts. The bluntly-tipped lobes are often reddish brown with a purplish or metallic glint, but they may have a greyish or bluish tinge; there is a network of slightly raised lines on the surface (1A). The plant is similar to *P. saxatilis*, which it replaces at high altitudes, and also to *P. sulcata* (p. 163), but *P. saxatilis* has rod-like outgrowths (isidia) on the surface, and *P. sulcata* has powdery reproductive structures (soredia) which develop along the network of ridges.

Parmelia prolixa is found in upland districts and also by the sea, closely attached to rock surfaces. It is dark brown and shining, and has rather narrow, crowded, overlapping lobes which are slightly turned under at the margins and faintly marked with wrinkles.

2 **Parmelia conspersa** grows on rocks in hilly districts. The long, greenish-yellow, faintly-wrinkled lobes broaden slightly at the tips. The spore-producing apothecia are conspicuous reddish-brown discs growing on the surface. In older plants, rod-like outgrowths (isidia), which are rather longer than those found in some other species of *Parmelia*, develop in the centre.

3 **Parmelia incurva** is uncommon except in the extreme north of Britain. It forms round, greenish-grey patches on rock. The lobes are narrow, crowded, and overlapping, and the centre of the plant is darker in colour and irregular and uneven. Whitish- or yellowish-grey, powdery reproductive structures (soredia) develop in globular heads on the surface (3A).

Parmelia mougeotii is similar to *P. incurva*, but the greenish-yellow, narrow-lobed, round patches are flatter and smoother. The pale yellow soredia are scattered on the surface and not in globular heads.

4 **Parmelia delisei**, which used to be called a variety of *P. prolixa*, grows in similar places but is a commoner plant. The pale olive-green lobes are narrow and crowded and slightly turned under at the margins.

5 **Parmelia fuliginosa**, sometimes regarded as a subspecies of *P. glabratula* (p. 160), grows closely attached to rocks, and has very dark brown or almost black lobes. The centre of the plant is covered with short, branched, black outgrowths called isidia (5A).

6 **Parmelia isidiotyla** forms brown, rather irregular patches on rocks in upland areas, and also on sandstone in the lowlands. Short outgrowths (isidia) grow in clusters on the surface, and give rise to pale-grey, powdery reproductive structures, called soredia, at their tips (6A).

7 **Menegazzia terebrata** grows on rocks and also sometimes on tree trunks, especially in the west of Britain. It forms neat and often quite large, closely attached, yellowish grey-green rosettes of slightly wavy lobes. The surface is perforated with neat round holes, almost as if the plant had been blasted at short range by a shot-gun loaded with dust-shot, Powdery reproductive structures (soredia) develop in small globular clusters, which are scattered all over the surface (7A).

8 **Placopsis gelida** colonizes hard rock surfaces on mountains, forming light pinkish-grey patches which are so closely attached to the rock that it is impossible to remove the plants without completely breaking them up. The margins consist of narrow, flat, radiating lobes with rather spreading, minutely notched tips. The centre is divided into small irregular areas by a series of fine cracks. The spore-producing apothecia are small reddish-brown to purplish-red discs with thick margins which occasionally appear on the surface. Circular patches of powdery reproductive structures (soredia) may sometimes be found also, but the most conspicuous features are the pale brown or flesh-coloured, wart-like outgrowths, called cephalodia, which contain the cells of a blue-green alga.

9 **Squamarina crassa** grows on soil in crevices in limestone rocks. The plant consists of thick, slightly overlapping 'scales' or squamules which are dark green when wet, but become brownish-green when dry, sometimes with a very fine powder or 'bloom' (pruina) on the margins. The undersides of the squamules are dark brown or almost black. The spore-producing apothecia are reddish-brown discs with a thin margin.

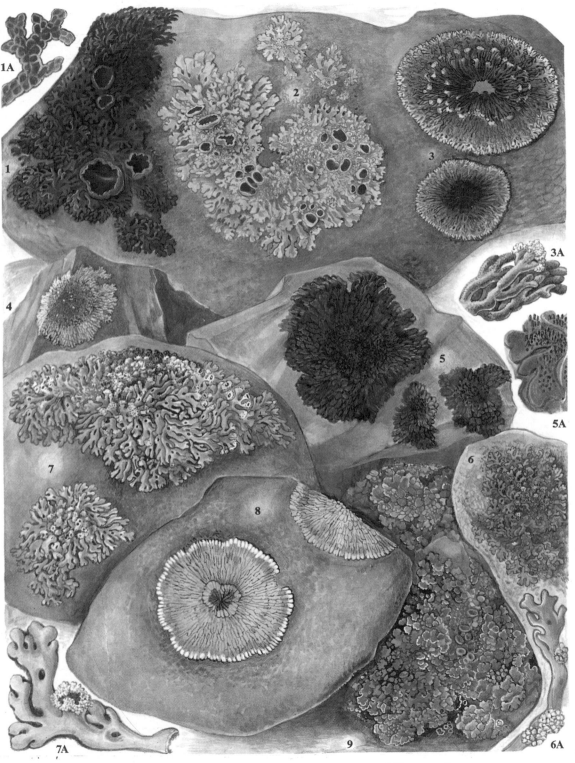

1 Parmelia Omphalodes   2 Parmelia Conspersa   3 Parmelia Incurva
4 Parmelia Delisei   5 Parmelia Fuliginosa   6 Parmelia Isidiotyla
7 Menegazzia Terebrata   8 Placopsis Gelida   9 Squamarina Crassa

1 **Pseudocyphellaria crocata** is a rather rare plant found in the western parts of Britain, most frequently in the highlands and islands of Scotland. It grows over and amongst mosses and liverworts on rocks and the trunks of trees. The greenish-brown upper surface is patterned with lines and clusters of bright yellow, powdery reproductive structures (soredia). The inner layers of the plant (medulla) are also bright yellow, and show as conspicuous spots through small holes (pseudo-cyphellae) scattered over the lower surface. *Sticta* species (p. 137) also have holes on the lower surface, but they are more elaborate in structure, with a distinct rim, and are called cyphellae.

2 **Pseudocyphellaria thouarsii** grows on mossy rocks and tree trunks in western Britain, especially in the Scottish highlands, though it is not a common plant. It is rather stiff and rigid, and the upper surface is brown with reddish or purplish tints when dry, but develops a distinctly green tinge when wet. It is covered with clusters of greyish-mauve reproductive structures (soredia), which develop also along the margins. Small openings (pseudocyphellae) are scattered on the lower surface.

3 **Peltigera horizontalis** is found in moist places in upland districts, growing on the ground or on rocks or tree trunks, often amongst mosses. The smooth upper surface is brown when dry, but develops a greenish tinge when wet, and there is a pattern of veins from which grow anchoring root-like structures (rhizines) on the brownish-white, hairy under surface. The plant is similar to *P. polydactyla* (p. 29), but the reddish-brown, oval, spore-producing apothecia are held horizontally rather than vertically.

4 **Peltigera aphthosa** grows on wet rocks on mountains, especially in Scotland. The upper surface is bright green when wet, but becomes brownish green on drying, and the whitish-brown, hairy underside has prominent veins and rhizines. Unlike *P. horizontalis* and the other species of *Peltigera* illustrated in this book (p. 29), the majority of the cells of algae which are present in this lichen are green and belong to a species of *Trebouxia*. The blue-green alga *Nostoc*, the alga present in the other species, is here restricted to dark granular outgrowths on the surface called cephalodia.

Peltigera venosa is a smaller plant, less common than *P. aphthosa*, but also with green algal cells. The cephalodia develop on the under surface, along the veins.

5 **Psoroma hypnorum** forms crusts of very small granular 'scales' or squamules, overgrowing mosses and liverworts on damp ground or rocks in mountain districts. It is here shown growing on *Hypnum cupressiforme* and *Frullania tamarisci* (p. 177). The spore-producing apothecia (5A) are reddish brown and have granular margins.

6 **Nephroma laevigatum** grows on mossy rocks and trees in the west and north of Britain. The upper surface is dark greenish-grey when wet but becomes brownish-green on drying, and the under surface is pale brown and finely wrinkled, but without hairs. The reddish-brown, spore-producing apothecia grow at the tips of the rather narrow lobes, and are fixed to the lower surface (6A). This unique position of the apothecia distinguishes the plant from species of *Peltigera* and *Parmelia* for which it might be mistaken.

7 **Leptogium burgessii** is found growing amongst mosses on rocks and trees, especially in Scotland and Wales. The numerous small lobes are greenish grey, slightly translucent, thin, and somewhat jelly-like in texture when wet, and are bluish grey and papery when dry. The underside is finely downy and paler in colour, and the margins have rounded teeth. The spore-producing apothecia (7A) have reddish-brown discs and numerous tiny 'leaflets' or squamules growing from their margins.

8 **Solorina saccata** grows on soil in the crevices of limestone rocks in upland districts. It is one of the very few lichens which are bright green when wet, and might be mistaken for a liverwort such as a species of *Pellia* (p. 100), but on drying it becomes a paler brownish-green with a trace of a very fine powder or 'bloom' (pruina) on the surface. The colour is due to a layer inside the plant of cells of a green alga belonging to a species of *Dactylococcus*. Below this layer there are groups of cells of the blue-green alga *Nostoc*. The spore-producing apothecia have dark-brown discs which are recessed slightly into the upper surface.

9 **Solorina crocea** is found on damp soil at high altitudes in the Scottish highlands. The smooth upper surface is green when wet, and reddish brown when dry. The undersurface and the interior of the plant is bright orange. Cells of a blue-green alga form a distinct layer beneath the layer of cells of a green alga. The large, dark-brown, spore-producing apothecia develop on, but not sunk into, the upper surface.

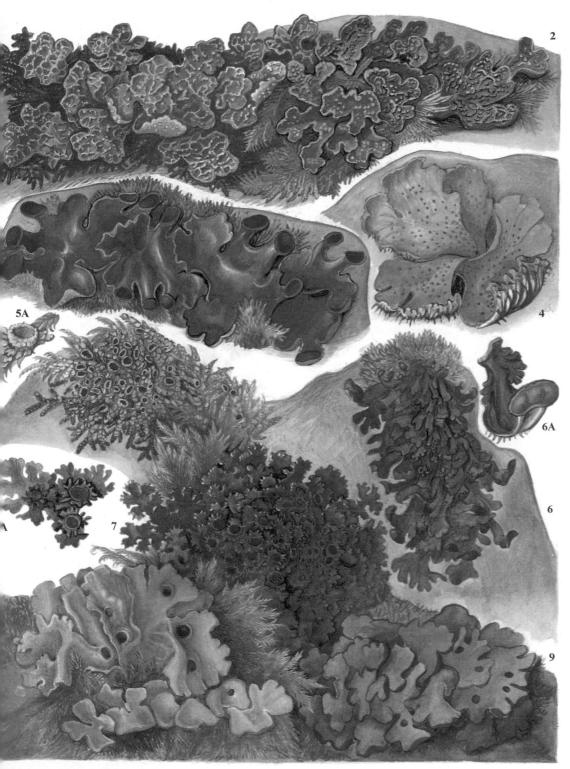

1 Pseudocyphellaria Crocata   2 Pseudocyphellaria Thouarsii   3 Peltigera Horozontalis
4 Peltigera Aphthosa   5 Psoroma Hypnorum   6 Nephroma Laevigatum
7 Leptogium Burgessii   8 Solorina Saccata   9 Solorina Crocea

65

Lichens are a large and successful group of plants in which each species consists of a fungus living in a very close association with an alga. The fungus forms the outer layer (cortex), while the inner layer (medulla) contains algal cells entangled in fungus threads. The algal cells are usually those of a green alga, often a species of *Trebouxia*, but sometimes they belong to a blue-green alga, nearly always a species of *Nostoc*. This close association, which produces a more elaborate and longer-lived plant than either partner can form alone, is called symbiosis. Lichens sometimes grow upright and are much branched, as in species of *Cetraria* (Nos. 1 and 2), but many of them look like a single flat, lobed leaf, as in the other species on this page. They may be attached to the surface on which they are growing in various ways, but species of *Umbilicaria* (Nos. 3 to 7) grow attached to rock by a single central point. A large number of lichens form crusts on rock (pp. 3, 77, 79, etc.) or on the bark of trees (pp. 169—173).

1 **Cetraria islandica** (Iceland Moss) grows amongst heather and other plants on mountains and moorlands, especially in Scotland and northern England. It is also found at a low altitude on a heath in Lincolnshire. The brown, branching fronds have a row of stiff, blunt spines along the margins, and small, scattered openings (pseudocyphellae) appearing as white dots on the under surface. An edible jelly can be made by boiling the plant, which should be thoroughly soaked before cooking to remove the bitter flavour.

2 **Cetraria nivalis** grows on mountain tops in Scotland. The branching fronds are cream coloured, tinged with yellow, and have a raised pattern of ridges, with hollows in between. The margins have a crisped appearance.

3 **Umbilicaria cylindrica** is common on mountain rocks in Scotland and Wales, and has rather thick, grey lobes, which are tinged pink on the under surface and fringed with robust, stiff hairs at the margins. The spore-producing apothecia (3A) have black, spirally-grooved discs.

4 **Umbilicaria pustulata** (*Tripe de roche*) grows on rocks and walls in the western parts of Britain. The brownish-grey upper surface is often tinged with green and has a 'pebbled' appearance, the raised 'bumps' being formed by corresponding hollows underneath. Spore-producing apothecia are rare, but there are usually short rod-like outgrowths (isidia) on the upper surface. The plant is edible if prepared in a similar manner to the edible seaweed *Porphyra* (p. 22), but it is a curiosity rather than a delicacy.

5 **Umbilicaria torrefacta** is found in upland districts the north and west of Britain. The upper surface dark grey or greyish brown, and the under surface grey or black. It is divided and perforated at the margins and has a lace-like appearance. The spore producing apothecia are small, and have black spirally-grooved discs.

6 **Umbilicaria polyrrhiza** is frequent in Scotland and northern England, and has a smooth, glossy, copper brown upper surface. The under surface is densely covered with black, branched hairs, which project outwards at the margins.

7 **Umbilicaria polyphylla** is common on mountain rock in Scotland and Wales, and has rather thin lobes with wavy margins. The upper surface is dark brown and smooth, and the under surface is dull black.

8 **Dermatocarpon miniatum** is common on damp rock especially in upland districts. It is attached by a single central point, like a species of *Umbilicaria*, but it has tiny, sunken, flask-shaped, spore-producing structures. The upper surface is light grey and, when dry, is covered with a very fine dust-like powder (pruina). The under surface is pale brown and smooth.

Dermatocarpon aquaticum is the only British lichen apart from some species forming crusts on rock which grows in water. It looks rather like a liverwort such as Pellia (p. 101), but is much tougher in texture and it turns brown without shrivelling when dried. grows on rocks in streams in upland districts.

1 Cetraria Islandica   2 Cetraria Nivalis
3 Umbilicaria Cylindrica   4 Umbilicaria Pustulata   5 Umbilicaria Torrefacta
6 Umbilicaria Polyrrhiza   7 Umbilicaria Polyphylla   8 Dermatocarpon Miniatum

For a general note on lichens see page 66.

1 **Sphaerophorus globosus** forms dense, pale-brown cushions, usually tinged with pink, on rocks and walls in upland areas. The lower parts of the branching stems are rather thick, hard, and cylindrical, and the upper, crowded branches are much smaller, flattened, and brittle. The spore-producing apothecia (1A) are globular, and when ripe they tear open irregularly at the top to expose the black spores within.

Sphaerophorus fragilis has grey, slender, repeatedly-forked stems, and the upper branches are not markedly different from the lower ones. It forms very compact cushions in similar places to *S. globosus*, and is equally common.

2 **Cladonia pityrea** is a common plant growing on soil, often amongst rocks. It forms a mat of small, thin, divided 'leaves' or squamules on the ground, which usually have fine, powdery, reproductive structures (soredia) on the surface. Slender stalks, which are covered with coarse granules and finer, powdery soredia, grow from the squamules. They are sparsely clothed with small, leaf-like scales, and produce irregular cups or whorls of short branches, on which grow brown, spore-producing apothecia (2A).

3 **Cladonia bellidiflora** is a handsome species which grows on peaty soil on mountains. The whole plant is bluish or greenish grey, except for the rather large, clustered, bright-red spore-producing apothecia. The sparingly-branched stems are covered with small scale-like 'leaves' or squamules, and grow from larger squamules at the base.

4 **Cladonia subcervicornis** forms cushions of 'leaves' or squamules, usually on well-drained soils. Each squamule (4A) is elongated and much divided, and is pale grey or greenish grey on the upper surface and white on the under surface. Brown, spore-producing apothecia are sometimes formed on rather large, swollen stalks.

Cladonia cervicornis grows in similar places to *C. subcervicornis*, but is not so common. The 'leaves' or squamules have less divided, dark greenish-grey upper surfaces. Brown, spore-producing apothecia

are sometimes formed on the margins of smal regular cups with short stalks. Other species o *Cladonia* are shown on pp. 29, 45, 51, 75, 85, and 175

5 **Thamnolia vermicularis** forms pale, greyish-white sparingly branched, creeping, hollow stalks wit pointed ends. It grows on the ground on mountain usually at high altitudes, and frequently amongst ma of the moss *Rhacomitrium lanuginosum* (p. 59). It doe not produce powdery reproductive structures (soredia or spores, and its only method of spreading is b pieces of stalk becoming broken off and continuin to grow elsewhere.

6 **Siphula ceratites** grows on one mountain only i Scotland, where it covers quite a large area of bar peaty ground with clusters of pale greyish-white upright stalks. It is quite a common plant in arct regions. Like *Thamnolia vermicularis*, its only metho of reproduction is by means of pieces of stalk breakir off and growing elsewhere.

7 **Stereocaulon vesuvianum** is common on rocks an walls in upland districts. The pale-grey, uprigh sparingly-branched stems have a tough and fibrou texture. They are covered with very tiny, rounde scales (phyllocladia), which have small, round lob at the margins (7A).

8 **Stereocaulon evolutum** forms loosely-attached cushior on rocks which usually have a thin covering of so The pale-grey branching stems are clothed with sma scales (phyllocladia). These are flattened and divide so that they look like tiny hands with the finge curled under (8A).

Stereocaulon coralloides is similar to *S. evolutum*, b grows firmly attached to bare rock by a centr 'holdfast'. The 'scales' (phyllocladia) on the stems a divided, but not flattened and hand-like.

Stereocaulon nanum grows on soil in moist, shad crevices. The very short and slender, sparingl branched stalks are completely covered with tin bluish-green, granular 'scales' (phyllocladia). Elev other species of *Stereocaulon* are to be found Britain, but all are uncommon or rare.

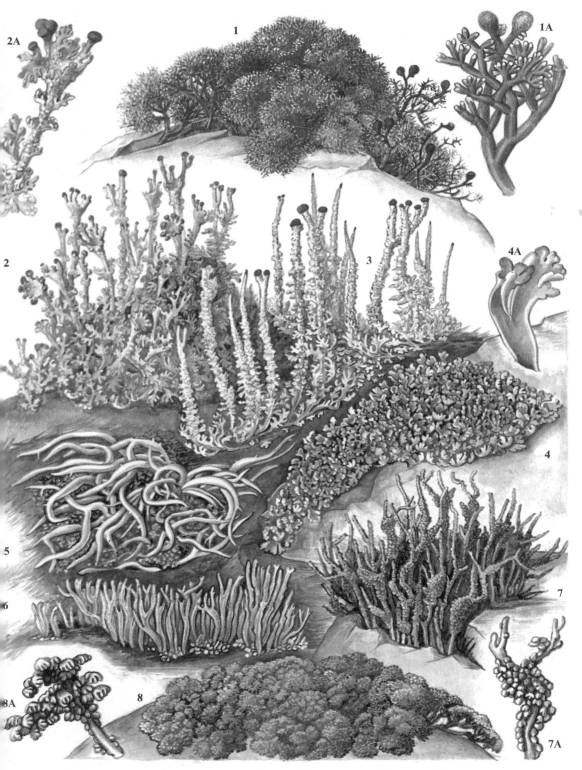

1 Sphaerophorus Globosus   2 Cladonia Pityrea   3 Cladonia Bellidiflora
4 Cladonia Subcervicornis   5 Thamnolia Vermicularis   6 Siphula Ceratites
7 Stereocaulon Vesuvianum   8 Stereocaulon Evolutum

For a general note on lichens see page 66.

1 **Ochrolechia tartarea** (Cudbear) forms thick grey crusts with irregular, rough, and lumpy surfaces on rocks and also on the bark of trees in upland districts. The spore-producing apothecia (1A) have hollowed, pale-brown discs and thick margins. A purple dye used to be made from this lichen by extracting it with ammonia.

2 **Pertusaria corallina** forms grey crusts on rocks and walls. There is a prominent white line along the margin, and well-developed plants are covered with closely-packed, pointed projections (2A). When the surface is touched with a drop of strong potassium hydroxide solution, it turns bright yellow and then orange.

P. pseudocorallina is a similar species which is common on rocks by the sea. The projections are sometimes branched, and the surface turns yellow and then blood red when touched with a drop of potassium hydroxide solution. Other species of *Pertusaria*, which grow on trees, are shown on p. 169.

3 **Diploschistes scruposus** is common on rocks and walls, sometimes spreading over mosses. It has a grey, cracked surface, and the spore-producing apothecia develop within small, round warts, their discs appearing as small, dark openings when ripe (3A). The apothecia are imbedded in warts, but singly, and not in groups of two or three as are those of *Pertusaria*.

4 **Solenopsora candicans** grows on limestone, forming greyish-white, circular patches that are cracked in the centre and lobed at the margins. The only other British lichen that has this growth form is *Buellia canescens* (p. 79), but this always has a bluish or greenish tinge and does not grow on limestone. The small, spore-producing apothecia have thin white margins and black discs covered with a very fine powder (pruina).

5 **Haematomma ventosum** forms extensive yellowish-green crusts on rocks in upland districts. The conspicuous spore-producing apothecia are the colour of bloodstains.

Haematomma coccineum is common on shaded rocks throughout Britain, forming soft white crusts tinged pale green and with pure white margins. The small spore-producing apothecia are the colour of fresh blood.

6 **Rhizocarpon geographicum** is a common and conspicuous species on hard rocks on mountains. The bright yellow-green upper layer of the plant develops on a very thin black under layer (hypothallus) which shows as a black line at the margins and through cracks in the surface. When several plants grow together touching one another, the black lines have the appearance of boundaries on a map. The small, black, spore-producing apothecia are sunk in the surface of the upper layer (6A).

7 **Rhizocarpon umbilicatum** grows on limestone, forming neat, round, greyish-white patches, with a pattern of cracks on the surface and small, rounded lobes at the margins. The small, black, spore-producing apothecia are sunk in little 'islands' on the surface (7A).

8 **Lecidea macrocarpa** is a very common species, forming cracked, grey crusts on stones and rocks. The large, black, spore-producing apothecia have a dark reddish tinge inside, which can be seen if they are cut open with a knife.

9 **Lecidea dicksonii** is frequent on rocks in upland districts. The rust-coloured upper layer develops on a very thin, black under layer (hypothallus) which shows as a black line at the margins and through cracks in the surface. The small, black, spore-producing apothecia are sunk in little 'islands' on the surface (9A). The plant looks rather like a species of *Rhizocarpon*, but when examined under the microscope the spores are found to be different. There are 170 species of *Lecidea* that grow in Britain, only a few of which can be shown in this book, here and on pp. 53, 79, 169, and 171.

1 OCHROLECHIA TARTAREA   2 PERTUSARIA CORALLINA   3 DIPLOSCHISTES SCRUPOSUS
4 SOLENOPSORA CANDICANS   5 HAEMATOMMA VENTOSUM   6 RHIZOCARPON GEOGRAPHICUM
7 RHIZOCARPON UMBILICATUM   8 LECIDEA MACROCARPA   9 LECIDEA DICKSONII

Walls provide a variety of habitats in which flowerless plants may grow — in crevices and joints between the stones, on the wall tops where small amounts of soil may accumulate, and on the stones themselves. Plants that avoid limestone are found on sandstone and brick, while those needing it can grow on the mortar in the joints. Before walls existed, all these plants grew only on and amongst rocks and boulders, and so are included in 'Uplands' here; though many can be found on walls anywhere. Wall ferns all prefer limestone, and are shown on a limestone wall on this page. Wall mosses are shown on p. 75. Lichens are shown growing on limestone on p. 77 and on brick and cement on p. 79.

1 **Asplenium viride** ('Green Spleenwort') grows in mountain districts in Scotland, northern England, and a few places in Wales. The fronds are divided into about 30 pairs of leaflets which have a few broad, round lobes at the margins. The leaflets are arranged alternately on slender green stalks, and do not fall off with age as do those of *A. trichomanes*. The spore cases develop, as in all species of *Asplenium*, in oblong clusters covered by papery protective scales (indusia) on the undersides of the fronds.

2 **Asplenium trichomanes** ('Maidenhair Spleenwort') is common throughout Britain. The fronds have up to 30 or 40 opposite pairs of leaflets with fine round-toothed margins, arranged on a wiry, shiny, purplish-black stalk. In a frond's second season of growth the leaflets fall off from the bottom upwards, eventually leaving the stalk quite bare. The spore cases develop on the undersides of the fronds in oblong clusters covered by papery protective scales (2A), which fall off before the spores are ripe.

3 **Asplenium adiantum-nigrum** ('Black Spleenwort') is quite a common plant, especially in the west of Britain. The fronds are narrowly triangular in outline, and have long, purplish-brown stalks clothed with small, bristle-like scales, especially at the base. They are divided into about 15 pairs of leaflets, which are further divided into short-stalked lobes, with lobed and toothed margins. The spore cases develop on the undersides of the fronds in oblong clusters protected at first by papery scales (indusia), but these fall off before the spores are ripe (3A). New fronds develop

quite late in the season, and are not fully unrolled until midsummer.

4 **Asplenium ruta-muraria** (Wall Rue) is common throughout Britain. The stalks of the fronds are branched twice, and end in wedge-shaped leaflets which are slightly toothed at the tips. The spore cases develop on the undersides of the fronds in oblong clusters covered by thin papery scales (indusia), with finely-toothed margins. A seaside species, *A. marinum*, is shown on p. 1.

5 **Cystopteris fragilis** ('Bladder Fern') is quite common in the north and west of Britain, but rare in the south east. The rather narrow and fairly long-stalked fronds are very variable in size and shape, but they are always fragile and delicate in texture. They are divided into about 15 pairs of main leaflets, each further divided into smaller leaflets, with deeply-lobed margins. The spore cases develop in groups on the undersides of the fronds, each covered with a delicate, domed scale, shaped like a hood (5A).

6 **Ceterach officinarum** (Rusty-back Fern) is quite a common plant in the south and west of Britain, but less common in the north. The thick, rather leathery, dark-green fronds are deeply lobed, but are not divided into leaflets. The undersides are densely covered with overlapping scales; these are silvery in young fronds, but change to light brown with age, and eventually become a rusty colour. The spore cases develop in clusters which have no special protective scale (indusium), but are concealed amongst the ordinary scales (6A).

1 ASPLENIUM VIRIDE   2 ASPLENIUM TRICHOMANES   3 ASPLENIUM ADIANTUM-NIGRUM
4 ASPLENIUM RUTA-MURARIA   5 CYSTOPTERIS FRAGILIS   6 CETERACH OFFICINARUM

1 **Bryum capillare** is a very common moss which forms compact cushions on wall tops and roofs. The oblong leaves are broadest in the middle and have well-developed nerves running into long protruding 'hair points' at the tips (1A). The leaves become spirally twisted to a marked degree when dry, and the 'hair points' are not silvery like those of *Tortula muralis* and *Grimmia pulvinata*. The spore capsules are bright green when young, but become brown when the spores are ripe. Their stalks are tinged red at the base.

2 **Bryum argenteum** grows on walls, roofs, and rocks, and in the cracks between paving stones, forming dark-green patches with a silvery sheen. It is one of the most widespread of all mosses, having been found on cliffs by the sea where fulmars nest, and by the roots of cactus plants in the Arizona desert. The oval leaves, which clasp the stems closely (2A), have sharp points and nerves that extend almost to the tip. Other species of *Bryum* are shown on pp. 58 and 83.

3 **Tortula muralis** is a very common moss on walls, where it forms small, neat cushions. The leaves have well-developed nerves, and long, silvery 'hair points'. The spore capsules grow upright on short stalks which are yellow when young, but become purplish red with age. A young capsule, shown in the centre of the tuft in the illustration, is still covered with a pointed 'hood' (calyptra). When the spores are ripe and the 'hood' and the lid beneath it have fallen off, the long, spirally-twisted, hair-like teeth surrounding the mouth of the capsule can be seen (3A). A related species, *T. ruraliformis*, is shown on p. 29.

4 **Cladonia fimbriata** is a common lichen on walls, and also on tree stumps and on the ground. It forms flat cushions of small greyish-green 'scales' or squamules, from which grow long-stalked cups shaped like wine glasses. The whole surface of both cups and stalks are covered with powdery reproductive structures (soredia) with a texture like that of flour. Other species of *Cladonia* are shown on pp. 29, 45, 51, 69, 85, and 175.

5 **Collema crispum** is a lichen with a soft and rather jelly-like texture. It has small, overlapping greenish or blackish-brown lobes, often with small scales (folioles) at the margins and short, rod-like outgrowths (isidia) in the centre. The spore-producing apothecia (5A) have dark-red discs and thin margins. As in all species of *Collema*, the alga in this lichen is a species of the blue-green alga *Nostoc* (*see* p. 66).

6 **Collema tenax** forms small, irregular, rather thick and swollen, greenish or blackish-brown masses, which have a soft and somewhat jelly-like texture. It is common on wall tops and also on the ground. The spore-producing apothecia have dark-red discs and thin margins. Another species, *C. furfuraceum*, is shown on p. 166.

7 **Grimmia pulvinata** is a common moss on roofs and wall tops, often on limestone, where it forms compact, domed cushions. The narrow, triangular-shaped leaves have well-developed nerves running into very long, wavy, silvery 'hair points'. The spore capsules are marked with eight lengthwise grooves and ridges, and have lids with pointed beaks. Their stalks are very much curved at first, the capsules being buried amongst the leaves until the stalks straighten as the spores ripen.

8 **Grimmia apocarpa** forms cushions on wall tops and boulders, often but not always on limestone. In the form shown here the leaves are tinged reddish-brown and have white 'hair points', but in other common forms they are deep green, and the 'hair points' are very short and inconspicuous. The spore capsules grow upright on very short stalks, so that they remain partly hidden in the leaves (8A). They have lids with curved beaks. Another species, *G. maritima*, is shown on p. 1.

9 **Camptothecium sericeum** is a moss growing in extensive mats on walls, boulders, and the base of trees. The repeatedly-branched stems are clothed with narrow triangular-shaped leaves that taper to fine points and have nerves extending almost to the tips. When dry the side branches curl up (9A). The whole plant has a very glossy, silky appearance. A related species *C. lutescens*, is shown on p. 43.

10 **Barbula convoluta** is a common moss on wall tops, cracks in pavement, and on bare ground, forming yellowish-green mats or cushions. It has long, oval leaves narrowed at their base and with short, sharp points at their tips. The nerves run almost as far as the points (10A).
B. unguiculata is a similar but rather larger plant which the nerves of the leaves run into longer and more protruding points.
B. fallax has triangular leaves which taper gradually towards the tips, and the nerves run the whole length

1A 1 3 3A

5A

4 5

8A

8

7

9A

9

2A 2 6 10

10A

1 BRYUM CAPILLARE   2 BRYUM ARGENTEUM   3 TORTULA MURALIS   4 CLADONIA FIMBRIATA

5 COLLEMA CRISPUM   6 COLLEMA TENAX   7 GRIMMIA PULVINATA

8 GRIMMIA APOCARPA   9 CAMPTOTHECIUM SERICEUM   10 BARBULA CONVOLUTA

There is a general note on lichens on page 66.

1 **Lecanora campestris** is common on rocks and walls, forming grey, granular crusts, bordered with a thin white line. The spore-producing apothecia (1A) have brown discs, usually tinged dark red, and wavy, grey margins.

2 **Lecanora calcarea** grows on limestone, forming cracked, granular crusts, which are grey in the centre, and chalky white towards the edges; they have a thin bluish-grey line along the margins. Spore-producing apothecia develop within the crust (2A), but burst through the surface before the spores become ripe. There are over 80 species of *Lecanora* that grow in Britain, only a few of which can be described in this book — here and on pp. 3, 7, 53, 77, 79, 171, and 173.

3 **Placynthium nigrum** grows commonly on limestone and cement in irregular patches. These consist of an upper layer of closely-packed, dull-black granules, growing on a very thin under layer (hypothallus). This under layer is black, tinged dark blue, like blue-black ink, and shows as a line at the margins and through cracks in the surface.

4 **Physcia adscendens** forms rather loosely-attached, light-grey rosettes on stone and on wood. The narrow, branching lobes are turned upwards towards the tips. Short, thick, rather bristle-like hairs grow from the margins, and greenish-grey, granular reproductive structures (soredia) develop in masses underneath the helmet-shaped ends of the lobes (4A).

5 **Physcia orbicularis** grows in flat, neat rosettes on rocks and trees. They are grey when dry, as in the plant on the left, but when wet become bright green in some plants, often called form *virella*, or greenish-brown in others. Powdery reproductive structures (soredia) develop in small flat patches on the surface. Other species of *Physcia* are shown on pp. 78, 165, and 169.

6 **Caloplaca heppiana** is common on limestone and cement, forming orange-yellow, closely-attached patches with lobed margins. The lobe ends (6A) have a rather swollen or 'domed' appearance. Orange, spore-producing apothecia develop in the centres of the rosettes.
C. aurantia is a similar, equally common species, but it is usually yellow rather than orange, and the ends of the lobes are flat and spreading.

7 **Caloplaca cirrochroa** is a rather rare plant which forms round, orange patches with lobed margins on shaded limestone rocks and walls. Bright yellow, powdery reproductive structures (soredia) grow in small clusters on the surface.

8 **Caloplaca citrina** grows in irregular yellow patches on limestone, cement, wood, and old leather on beaches. Like all yellow and orange species of *Caloplaca*, it changes to a deep purple colour when touched with a drop of potassium hydroxide solution. Small yellow or orange-yellow, spore-producing apothecia are often produced. Other seaside species of *Caloplaca* are described on p. 2.

9 **Xanthoria aureola** is common on rocks and walls. It is very similar to *X. parietina* (pp. 3 and 169), but the surfaces of the deep-yellow or orange-yellow lobes are covered with short, thick, rod-like outgrowths with rounded tips. Quite large plants are often found without apothecia, but when these are developed, as shown here, they are comparatively large and the same colour as the rest of the plant.

10 **Verrucaria nigrescens** is common on limestone and cement, growing in irregular patches, consisting of a thin, brownish-black upper layer growing on an even thinner, black under layer (hypothallus). Spores are produced in small, flask-shaped structures which appear as tiny black dots on the surface. Other species of *Verrucaria* are shown on p. 7.

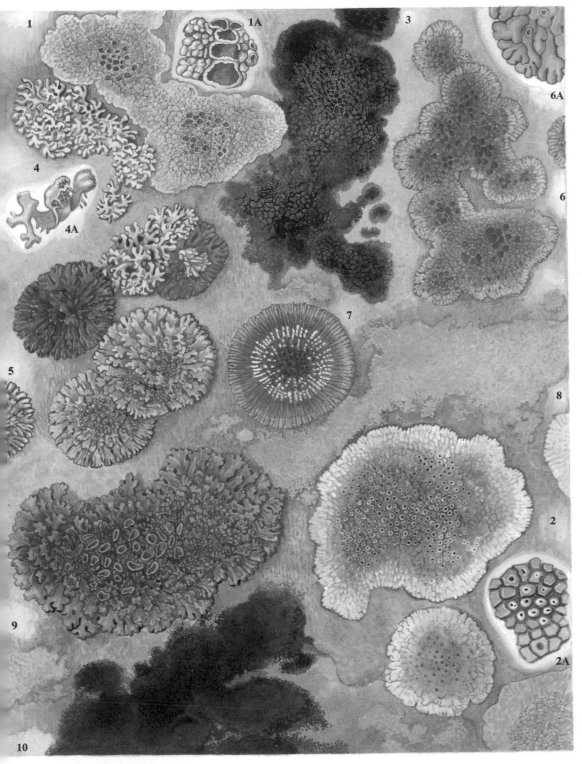

1 Lecanora Campestris   2 Lecanora Calcarea   3 Placynthium Nigrum

Physcia Adscendens   5 Physcia Orbicularis   6 Caloplaca Heppiana   7 Caloplaca Cirrochroa

8 Caloplaca Citrina   9 Xanthoria Aureola   10 Verrucaria Nigrescens

For a general note on lichens see page 66.

1 **Lecidea coarctata** is common on damp rocks throughout Britain. It flourishes also on bricks, as shown here, and is often to be found on wall tops, even in towns, where other lichens are scarce. It is a variable plant, forming crusts which may be smooth and greenish white, cracked and whitish grey, or consist of very tiny pinkish-grey 'scales' or squamules. The small, spore-producing apothecia are reddish brown when young, but become darker brown or almost black with age (1A).

2 **Lecidea lucida** grows as a bright greenish-yellow powder in thin but extensive patches on moist bricks and rocks; it can flourish in partial shade. The spore-producing apothecia, which are sometimes to be seen, are very small and the same colour as the rest of the plant. *L. lucida* can be confused with *Caloplaca citrina* (p. 77) or *Candelariella vitellina*, both of which will grow only on bricks which have become impregnated with lime from the mortar.

3 **Buellia canescens** is very common on rocks and walls and can tolerate a certain amount of shade. It grows equally commonly on trees, and is shown on bark on p. 169. It forms smooth, very closely-attached patches, with narrow lobes at the margins and often with scattered patches of powdery reproductive structures (soredia) in the centre. When dry, the plant is white or greyish white with a faint bluish tinge, which either becomes intensified on wetting or, as shown here, changes to green.

4 **Physcia grisea** is found on walls, rocks, and trees, attached by short threads which grow from the underside and are white with black tips. The plant is light grey or slightly brownish grey, and is almost always covered, at least near the tips of the lobes, with a very fine white powder (pruina). The colour develops a greenish tinge when the plant is wetted. Groups of grey, granular, powdery, reproductive structures (soredia) grow along the margins and sometimes on the surface as well; but apothecia are rare.

5 **Physcia caesia** is common on walls and rocks, but does not usually grow on trees. The grey, narrow, branching lobes are speckled with scattered, very tiny openings in the surface (pseudocyphellae), and do not change colour when wetted. Whitish-grey, granular, powdery soredia develop in conspicuous clusters on the surface, but apothecia are rare. Other species of *Physcia* are shown on pp. 77, 165, and 169.

6 **Ochrolechia parella** grows on walls and rocks and forms thick grey crusts with a rough surface and a prominent white border at the margins. The apothecia are usually numerous and quite large, with pale pinkish-brown discs and thick, well-developed margins. A related species, *O. tartarea*, is shown on p. 71.

7 **Lecanora dispersa**. Although far from being the most conspicuous, this is the commonest lichen in Britain. It grows abundantly on limestone in all parts of the country and at all altitudes, and also on concrete, cement, and mortar everywhere, even in the centres of large cities where no other lichens are found. A form *hageni* also grows on wood. The plant forms a white or greyish-white crust, and has very numerous apothecia (7A), which are often densely crowded together.

8 **Lecanora muralis** forms conspicuous rosettes on concrete and cement, as well as on limestone rocks; it is frequently found on asbestos roofs. It has brownish-green, crowded, narrow, very closely-attached lobes at the margins, and numerous brown apothecia with prominent margins in the centre.

9 **Candelariella vitellina** grows on mortar and cement, as shown here, and on rocks, forming orange-yellow, granular crusts. Unlike *Caloplaca citrina* (p. 77) it does not change to a deep purple when touched with a drop of potassium hydroxide solution. Small, yellow, spore-producing apothecia (9A) are frequently produced.
C. aurella is similar, but consists of scattered granules rather than a continuous crust, and produces numerous apothecia.
C. medians grows in pale orange rosettes on limestone. It looks rather similar to *Caloplaca heppiana* (p. 77), but does not turn purple with potassium hydroxide solution.

10 **Rinodina exigua** is common on walls and rocks, forming small, scattered, dark brownish-grey patches with numerous similarly-coloured apothecia.

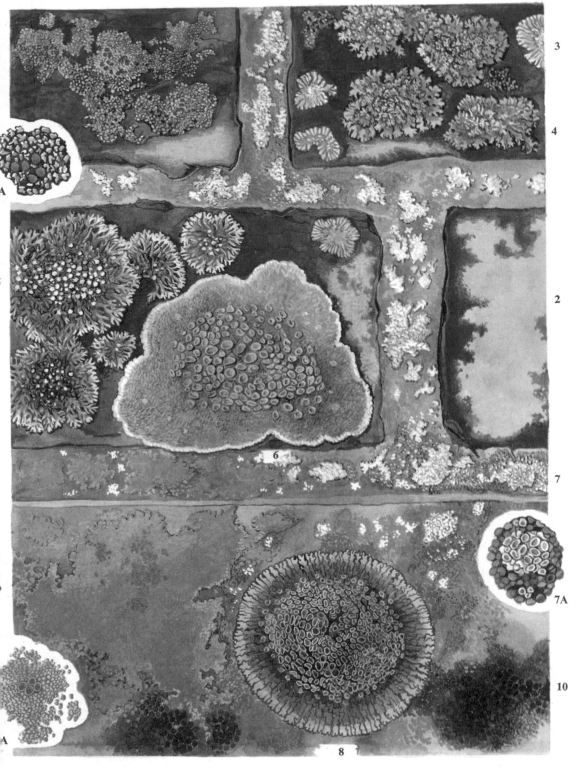

1 Lecidea Coarctata   2 Lecidea Lucida   3 Buellia Canescens   4 Physcia Grisea
5 Physcia Caesia   6 Ochrolechia Parella   7 Lecanora Dispersa
8 Lecanora Muralis   9 Candelariella Vitellina   10 Rinodina Exigua

Moors develop on peaty soils at fairly high altitudes in western and northern Britain. Usually the dominant flowering plant is heather, sometimes with abundant bilberry. Two common communities dominated by grasses are 'purple moor', with the grass *Molinia caerulea*, and 'white moor' with mat grass (*Nardus stricta*). The ground is normally wetter than on heaths, but some of the flowerless plants described here may grow in both types of habitat. Some bog plants, ordinarily growing in very wet conditions, are also found on moors.

1 **Polytrichum formosum** is a common moss on mildly acid soils, and is found in woods and on hedge banks as well as on heaths and moors. As in all species of *Polytrichum*, the base of the leaves sheathes the stems, and they have a series of vertical plates running lengthwise along their upper surface, which makes them opaque. They have toothed margins, as in *P. commune*, but they do not spread so widely away from the stem. The spore capsules, when young, are covered by loosely fitting, very hairy 'caps' (calyptrae) and are usually 5- or 6-sided (1A).

2 **Polytrichum commune** ('Hair Moss') is the largest British moss, apart from some wholly aquatic species such as *Fontinalis antipyretica* (p. 97). It is common on wet moors with highly acid soils, on 'white moor', and amongst *Rhacomitrium lanuginosum* (p. 59) on mountains. The leaves have toothed margins and spread widely from the stems. The spore capsules are 4-sided and have a conspicuous swelling (apophysis) at the base (2A). The lid has a shorter beak than that of *P. formosum*. The hairy, golden-yellow 'cap' (calyptra) is very long and handsome.

3 **Polytrichum juniperinum** grows amongst heather and grasses on rather dry, acid soils on moors and heaths, forming extensive turfs. The margins of the leaves are not toothed, but there are a few small teeth on the pointed tips. The spore capsules (3A) are 4-sided and have bright red lids with short beaks. In the illustration, young capsules completely enclosed by yellow, hairy 'caps' (calyptrae) are shown at the bottom right; above, on the left, the stalks have elongated, and on the right the capsules are ripe and the calyptrae have

fallen off. Male plants are conspicuous because they have a rosette of orange or red tinted leaves at the tips of the stems, as shown at the bottom left. *P. piliferum* (p. 29) is similar, but the tips of the leaves are greyish white instead of tinged reddish brown, as in *P. juniperinum*.

4 **Polytrichum aloides** forms open turfs on disturbed ground on moorlands, beside paths, and in quarries, and is often found in shaded situations. The leaves taper rather abruptly at the tips, but do not have long bristle-like points, and the margins are toothed. When young, the spore capsules are covered with brown, very hairy 'caps' (calyptrae), and when ripe (4A), they are urn shaped, without flat sides, and are held upright on red stalks. The male plant (4B) has a rosette of reddish-brown leaves at the top of the stem.

Polytrichum urnigerum is similar to *P. aloides* and grows in the same kinds of places, but it forms taller turfs. The leaves are bluish green rather than dark green, and are narrower, with very sharply-toothed margins.

5 **Neottiella rutilans** is a cup fungus (*see* p. 150), quite commonly found in autumn and winter growing on sandy soils amongst turfs of species of *Polytrichum*. The upper, spore-producing surface of the short-stalked cup is yellow, usually tinged reddish orange, and the under surface is covered with a dense coat of white, downy hairs.

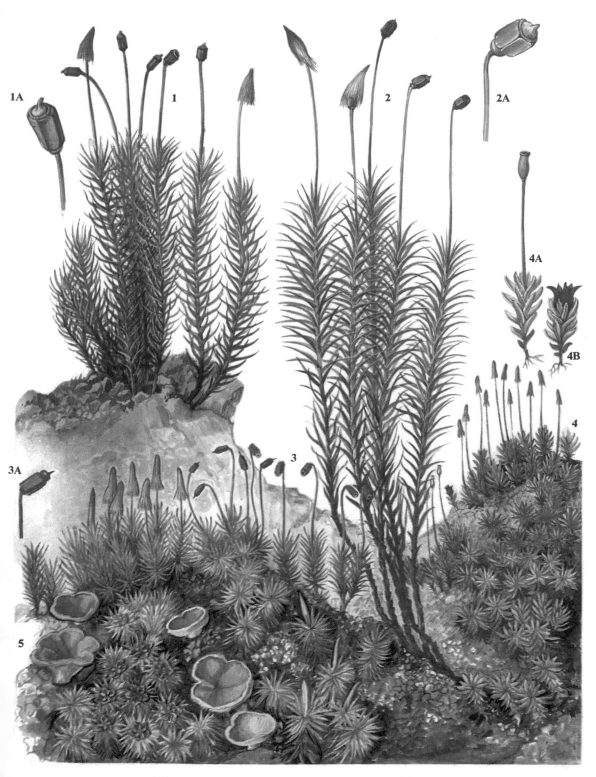

1A   1   2   2A

4A

4B

3A

3   4

5

1 POLYTRICHUM FORMOSUM   2 POLYTRICHUM COMMUNE
3 POLYTRICHUM JUNIPERINUM   4 POLYTRICHUM ALOIDES
5 NEOTTIELLA RUTILANS

1 **Calypogeia meylanii** is a liverwort growing on peaty and sandy soils and on sandstone rocks. The oval leaves grow in two ranks from slender stems. Each leaf overlaps slightly the one in front of it, in contrast to most British leafy liverworts in which the front edge of each leaf is concealed beneath the leaf in front. The underleaves are deeply divided into two lobes (1A). Other species of *Calypogeia* are described on p. 92.

2 **Nardia scalaris** is a common liverwort on moist, gritty soils throughout Britain. The short, almost unbranched stems form quite extensive yellowish-green patches, sometimes tinged reddish brown, and are clothed with round, hollowed leaves growing in two ranks (2A). The underleaves are small, narrow, and pointed. The plants are anchored to the soil by numerous fine white hairs.

3 **Bryum alpinum** is a moss of moors and mountains, growing on moist rocks and stony ground. The tall, sparingly-branched stems are clothed with reddish- or purplish-brown, glossy leaves. Each leaf is narrow and oblong, with incurved margins and a well-marked nerve running right to the pointed tip. Other smaller species of *Bryum* are shown on p. 75.

4 **Breutelia chrysocoma** is a moss of moist places on moors and mountains, mainly in the west of Britain. The red, upright stems are sparingly branched, and the spreading leaves give them a 'bottle brush' appearance. They are clothed with a felt of reddish-brown hairs (4A). Each leaf is broadest just above the base and tapers to a long fine tip. The margins are finely toothed, and the nerve runs almost the whole length of the leaf.

5 **Dicranum fuscescens** is a moss which grows in tufts on peat and rock in upland areas. The dull-green leaves are turned to one side and become markedly blackened towards the base of the plant. When dry, they are much twisted and curled, as shown in the smaller clump in the illustration. Each leaf tapers to a long, fine, curved point, which is covered with very small, spiny teeth. The nerve is fairly broad and runs the whole length. Other species of *Dicranum* are described on p. 182.

6 **Campylopus flexuosus** is a common moss on moorlands and in peaty places in woods, growing in compact tufts or cushions. The lower parts of the stems are densely covered with short, rather pale reddish-brown hairs. The leaves taper into lightly-toothed points, and have broad nerves running their whole length. When dry they have a wavy, curled appearance (6A). The spore capsules are oval and grow on curved stalks.

7 **Campylopus atrovirens** grows in wet places in upland areas in the north and west of Britain. It forms extensive turfs which have a dark-green or almost black, velvety appearance. The leaves taper into long, strongly-toothed points, which rather easily become broken off. The nerves occupy half the width of the leaves.

8 **Campylopus pyriformis** forms short turfs on peaty soils, often of considerable extent. The leaves (8A) have very broad nerves, which run into long, fine, channelled points, with toothed edges. They readily fall off and sometimes grow into new plants. The spore capsules are pear-shaped and grow on curved stalks.

1 Calypogeia Meylanii   2 Nardia Scalaris   3 Bryum Alpinum
4 Breutelia Chrysocoma   5 Dicranum Fuscescens
6 Campylopus Flexuosus   7 Campylopus Atrovirens   8 Campylopus Pyriformis

1 **Cladonia macilenta** is a common lichen on peaty soil on moors and on tree stumps, forming a rather sparse crust of bluish-grey 'scales' or squamules. Unbranched or very sparingly branched, hollow stalks grow upright from the squamules. These are pale grey, usually tinged distinctly bluish-grey, at least at the base, and they taper towards the tips, where small, red, spore-producing apothecia develop. The upper parts are covered with grey, powdery, reproductive structures (soredia).

Cladonia polydactyla is similar to *C. macilenta* and grows in the same kinds of places, but the stalks have narrow cups at their tips, often with shorter stalks growing from their margins. The upper parts, and sometimes the 'scales' or squamules at the base as well, are covered with fine, powdery, reproductive structures (soredia).

2 **Cladonia squamosa** grows on poor soils, forming a layer of rather small 'scales' or squamules on the ground. From these grow upright, hollow stalks, usually ending in shallow cups with irregular outgrowths from the margins, but sometimes sparingly branched and pointed at the tips. The whole plant is greenish grey, distinctly tinged with brown, and the outer layers (cortex) are partly peeled away, forming small, projecting, almost leaf-like outgrowths. The brown spore-producing apothecia grow on short stalks from the margins of the cups.

3 **Cladonia coccifera** is common on peaty soils and has wide, regular cups with distinct stalks. The 'scales' or squamules at the base are small, and the granular, warted surface is pale grey, usually tinged yellow. Large, red, spore-producing apothecia develop on the margins of the cups (3A).

4 **Cladonia strepsilis** grows in extensive patches or cushions on moors and heaths. The rather thin 'scales' or squamules are brownish green on the upper surface, greyish white on the under surface, and deeply cut at the margins. Brown, spore-producing apothecia develop at the tips of short, much-branched stalks growing from the squamules (4A). The plant contains a unique substance, strepsilin, which causes it to turn bright green if it is touched with a drop of domestic bleaching solution.

5 **Cladonia floerkeana** is common on peaty soil, forming a rather sparse crust of small, greenish-grey 'scales' or squamules. Short, unbranched or sparingly-branched stalks, with conspicuous, red, spore-producing apothecia at their tips, grow from the squamules. The stalks have a rough, granular surface, and are pale grey, never tinged bluish green, but sometimes distinctly charcoal grey. Powdery reproductive structures (soredia) develop in patches on the surface. Other species of *Cladonia* are shown on pp. 29, 45, 51, 69, 75, and 175.

6 **Coriscium viride** grows on moist peat and consists of beautiful bluish-green 'scales' or squamules, with delicate upturned edges. The under surface is white and smooth. No spore-producing structures have ever been found on this lichen, but it is very frequently associated with the fungus *Omphalina ericitorum*, and it is possible that the fungus in the lichen is this species (*see* p. 66).

7 **Omphalina ericitorum** is found on wet, peaty ground in summer. The top of the cap is hollowed and becomes funnel shaped with age; it is pale yellowish- or faintly pinkish-brown. This fungus is often, or perhaps always, associated with the lichen *Coriscium viride*. In a similar manner, the related species *O. sphagnicola* is associated with the simple lichen *Botrydina vulgaris* (p. 88).

1 Cladonia Macilenta   2 Cladonia Squamosa   3 Cladonia Coccifera
4 Cladonia Strepsilis   5 Cladonia Floerkeana
6 Coriscium Viride   7 Omphalina Ericitorum

BOG MOSSES

Extensive areas of north and west Britain, where rainfall is high, are covered with 'blanket bog', characterized by stagnant water and dominated by species of *Sphagnum*. The leaves of these plants are largely composed of hollow cells with small pores in their walls, so that they absorb and hold many times their own weight of water. They grow in dense masses, their lower parts decaying slowly to form deposits of peat, which frequently accumulate to considerable depths. In places where there is marshy ground round springs and beside streams, even in areas where the rainfall is not very high, 'valley bog' may develop. This differs from blanket bog in that there is always a current of water through it, and it is usually richer in mineral salts. *Sphagnum* species with a higher mineral requirement flourish in it. Another similar habitat, 'raised bog', is described on p. 88.

1 **Sphagnum palustre** forms pale green mats, usually in the drier parts of bogs that are only moderately acid. The plants are robust and have whorls of swollen branches in well-marked rosettes. As can be seen in the enlarged detail (1B), the large stem leaves have broad bases and become even broader towards the tips, while the branch leaves are oval, with the tips hooded over. The spore capsules (1A), which are like those of other species of *Sphagnum*, grow on short stalks and are spherical, with a small circular lid. As they ripen, the walls shrink so that the capsule becomes cylindrical and eventually develops a slight 'waist'. This shrinkage compresses the air within until the lid is blown off violently with an explosive sound, and the spores are scattered by the outrushing air.

2 **Sphagnum teres** grows in fens or in only slightly acid bogs, forming bright green mats of rather slender plants, with branches spreading in star-like whorls. The stem leaves are short and as wide at the tip as at the base; the branch leaves are narrow and oval, and become folded outwards when dry.

3 **Sphagnum papillosum** is the commonest species of *Sphagnum* in the drier parts of blanket bog, and also of valley bogs. The plants form pale brownish-yellow hummocks and are robust, with whorls of short, blunt-tipped branches forming compact rosettes. As in *S. palustre*, the leaves on the main stems have broad bases and widen still more towards the tips, and the branch leaves are oval, with the tips hooded over.

4 **Sphagnum pulchrum** is commonest in the south of England, and grows in valley bogs and in wet hollows in 'raised bogs' (p. 88), where it forms yellowish-green mats of robust plants which are often tinged with orange-brown. The stem leaves are small, oval, and boat-shaped, while those on the branches are broader and longer, and narrowed to a short point.

5 **Sphagnum compactum** is found in bogs which are only slightly acid, and in wet places on heaths. The plants are usually short with upwardly-growing branches, and form dense, grey-green cushions, often tinged orange brown, which are rather like those of *Leucobryum glaucum* (p. 181). The triangular stem leaves are minute, but the much longer, overlapping, branch leaves have parallel sides and inrolled margins, and are hooded over at the tips.

6 **Sphagnum subsecundum** var. **auriculatum** forms extensive yellow-green mats which may be tinged orange brown or purple in valley bogs and in wet places in woods. The usually rather slender plants have rosettes of branches which are curved to one side at the tips. The stem leaves are long, usually parallel-sided, and rounded at the tips, while the branch leaves are oval with pointed tips, and are curved to one side.

7 **Sphagnum subsecundum** var. **inundatum** grows in pools in bogs, forming loose masses of long stems, with rather widely-spaced whorls of branches. The stem leaves are large with rounded tips, and the branch leaves are long and pointed.

1 SPHAGNUM PALUSTRE   2 SPHAGNUM TERES   3 SPHAGNUM PAPILLOSUM
4 SPHAGNUM PULCHRUM   5 SPHAGNUM COMPACTUM
6 SPHAGNUM SUBSECUNDUM *var.* AURICULATUM   7 SPHAGNUM SUBSECUNDUM *var.* INUNDATUM

Bogs are usually very deficient in mineral salts, and the water is always acid. These conditions are especially suitable for the growth of species of *Sphagnum* (*see* p. 86). Marshy places which are not acid, called fen, are mainly found in lowland areas. Sometimes fen becomes choked with vegetation, and acid conditions develop at the top, so that species of Sphagnum can grow on the surface, forming 'raised bog'. Two other kinds of bog are described on p. 86.

1   **Sphagnum plumulosum** is the commonest species of Sphagnum in both blanket bog (p. 86) and in raised bogs in many places in the north and west of Britain. It forms large, brownish-green hummocks, usually tinged with purple or pink. The branches grow in well-marked rosettes, and have pointed tips which become bright and glossy when dry. The stem leaves are triangular and pointed, and the branch leaves are long, narrow, and oval, with the margins incurved at the pointed tips.

2   **Sphagnum cuspidatum** grows in pools and hollows in blanket bog (p. 86) and in very wet places on moorlands, forming pale yellowish-green, loose mats. The slender stems are very long, and the branches have a delicate feather-like appearance. The stem leaves are triangular and longer than they are broad, while the branch leaves are very long and narrow and have wavy margins when they are dry.

3   **Sphagnum fimbriatum** is common in lowland districts, where it forms dark-green mats in partly shaded places in fens and in open, boggy woodlands. The plants are slender, with delicate branches in well-defined rosettes. The stem leaves have broad tips fringed with hairs, and the branch leaves are long and pointed.

4   **Sphagnum rubellum** grows in clusters of slender, delicate, crimson-red plants on the tops of hummocks in acid bogs; the main part of each hummock usually consists of *S. plumulosum*, often with *S. cuspidatum* in the wettest position at the base. The stem leaves have blunt tips, and the branch leaves are long and pointed and turned slightly to one side when dry.

5   **Sphagnum recurvum** forms extensive mats in the wetter parts of valley bogs and in open, marshy woodlands. It is usually dark green, but it may be quite pale. It is similar to *S. cuspidatum*, but the stem leaves are short, triangular, and broader than they are long. The stem leaves are long and narrow and turned back at their tips when dry.

6   **Lepidozia setacea** is a liverwort which forms compact, dark olive-green patches of upright, thread-like stems. It usually grows amongst *Sphagnum* in bogs, but sometimes with other mosses on wet moorlands. The leaves (6A) grow in three ranks, and are deeply divided into three or four very narrow lobes. Another species of *Lepidozia*, which grows on damp wood, is shown on p. 175.

7   **Odontoschisma sphagni** is a liverwort which grows amongst *Sphagnum* in valley bogs and raised bogs, but not usually in blanket bogs (p. 86). The green leaves (7A) are frequently mottled with reddish brown. They overlap one another slightly and grow in two ranks from slender, thread-like stems.

8   **Omphalina sphagnicola** is a fungus found in bogs in late summer and autumn. The top of the cap is hollowed and becomes funnel-shaped with age; it is dark brown, sometimes with an olive tinge. The gills are greyish brown and run down the stalk. This fungus is often, perhaps always, a member of a close association between several quite different plants. Near the base of the stalk, small green spheres the size of mustard seeds are found; these consist of groups of cells of a green alga surrounded by the threads of a fungus, forming a simple lichen called *Botrydina vulgaris* (*see* p. 66). Both the lichen and the fungus always grow with *Sphagnum*. Other species of *Omphalina* are shown on pp. 47 and 85.

1 SPHAGNUM PLUMULOSUM   2 SPHAGNUM CUSPIDATUM   3 SPHAGNUM FIMBRIATUM

4 SPHAGNUM RUBELLUM   5 SPHAGNUM RECURVUM

6 LEPIDOZIA SETACEA   7 ODONTOSCHISMA SPHAGNI   8 OMPHALINA SPHAGNICOLA

1 **Drepanocladus revolvens** forms yellowish-green, glossy tufts, usually tinged orange or purplish red. This moss grows in pools and bogs on peat, especially in the north and west of Britain. The irregularly-branched stems are densely clothed with leaves which are very markedly turned to one side, each leaf being curved almost into a circle. The leaves have nerves running for about three-quarters of their length, and they taper to a point; the margins are not toothed.

2 **Drepanocladus aduncus** is found in pools and ditches throughout Britain, although it avoids waters which are acid. It forms loose, rather dark-green wefts of somewhat irregularly-branched stems. The leaves are narrowly triangular, usually with nerves that extend for half their length or slightly more, but which are sometimes much shorter; the margins are not toothed.

Drepanocladus fluitans is similar to *D. aduncus*, but prefers slightly acid conditions. The tips of the leaves are toothed and turned to one side.

3 **Drepanocladus uncinatus** grows in pale golden-green or yellowish-olive mats on wet rocks and boggy ground in upland areas. The main stems have regularly-arranged, short side branches, and the long, narrow leaves are turned to one side, each leaf being strongly curved into a sickle shape (3A). The leaves have narrow nerves running for about three-quarters of their length, and they taper to long, very fine and distinctly-toothed tips; their surfaces are creased by several lengthwise folds. The orange-brown spore capsules are curved, as in other species of *Drepanocladus*, and grow on long, similarly-coloured stalks.

4 **Acrocladium sarmentosum** forms loose, deep purplish-red tufts, often with patches of green and orange. This moss is found in mountain bogs and streams. The rather irregularly-branched stems are clothed with pointed, oval leaves which are narrowed at the base and hooded at the tips (4A). The nerves run almost the whole length of the leaves. Other species of *Acrocladium* are shown on pp. 43 and 175.

5 **Aulacomnium palustre** forms yellowish-green cushions in bogs and on wet moorland. The stems are covered with a matted coat of reddish-brown hairs (rhizoids). The oblong leaves have pointed tips and nerves which run almost the whole of their length. A related species of moss, *A. androgynum*, is shown on p. 175.

6 **Scorpidium scorpioides** is a moss growing on wet peat, often forming extensive mats of long stems. The oval leaves are broad, hollowed, and clasp the stem with the tips turned slightly to one side in such a way as to give the shoots a somewhat swollen appearance (6A). The surfaces of the leaves are wrinkled, but the margins are not toothed, and the nerves are very short, forked structures.

7 **Calypogeia muelleriana** is a liverwort which forms dark-green mats in marshes, on damp rocks, and in wet places generally throughout Britain, although it avoids chalk and limestone. The main leaves, which grow in two rows from rigid stems, are a little wider than long and rounded at the tips. Each leaf overlaps slightly the one in front of it. The underleaves are two or three times as wide as the stems, the upper third of each being divided into two rounded lobes. Other species of *Calypogeia* are described on pp. 83 and 93.

8 **Pleurozia purpurea** is a common liverwort in western Scotland and Ireland, where it forms purplish-red tufts, sometimes tinged with reddish yellow, in wet, peaty moorland. Upright shoots grow from branched creeping stems; they are clothed with leaves (8A), each of which has a large lobe with a coarsely-toothed tip, and a smaller lobe curved round on itself to form a water-holding 'pitcher'. There is a little flap at the bottom of the 'pitcher', which allows water to enter, but prevents it running out again. It is possible that microscopic animals become trapped in the 'pitchers' and, when they die and decay, supply some nutriment to the plant.

1 Drepanocladus Revolvens   2 Drepanocladus Aduncus   3 Drepanocladus Uncinatus
4 Acrocladium Sarmentosum   5 Aulacomnium Palustre   6 Scorpidium Scorpioides
7 Calypogeia Muelleriana   8 Pleurozia Purpurea

The CHAROPHYTA is a small and ancient group of plants which is not closely related to any other group. It includes species of *Chara* (Nos. 5 and 6) and *Nitella*, which is similar except that the branchlets are forked at the tips. They all grow in fresh water and usually have a distinct and not very pleasant smell, rather like rotting tomatoes. Some species become encrusted with a deposit of calcium carbonate, which has led to their sometimes being called 'Stoneworts'.

1 **Cephalozia bicuspidata** is a common liverwort which grows in *Sphagnum* bogs, in wet places on soil, and on rotting timber in woodlands. It is a tiny plant with thread-like stems and two ranks of pale-green leaves which are deeply divided into two pointed lobes. The spore capsules develop within protective tubular sheaths (1A).
C. connivens is similar and grows in the same kinds of situation, but the tips of the lobes of the leaves are turned inwards like the jaws of pincers.

2 **Calypogeia fissa** is found in *Sphagnum* bogs, but is also a common liverwort on wet sandstone rocks and sandy soil in shady places. The leaves are oval with deep notches at the tips, and they grow in two ranks from slender stems. Each leaf slightly overlaps the one in front of it, whereas in most other British leafy liverworts the front edge of each leaf is concealed beneath the leaf in front. The plant also has under-leaves, each deeply divided into two lobes with coarsely-toothed margins (2A). Small granular reproductive structures (gemmae) grow in clusters at the tips of thread-like, minutely-leaved branches.
C. trichomanis, which is very similar in appearance, grows on sandy and peaty soils and on sandstone rocks, often at high altitudes. The oval leaves are rounded at the tips or only very slightly notched, and the under-leaves are less deeply divided than those of *C. fissa*, and are not toothed. Other species of *Calypogeia* are shown on pp. 83 and 91.

3 **Riccardia pinguis** is a common liverwort in wet places, which looks rather like a poorly-developed species of *Pellia* (p. 101); but it is smaller, with narrow lobes which have no midrib and are rounded at the tips rather than notched. Short, rather irregular branches, produced at intervals, grow out at right angles.
R. multifida, which grows in similar places, especially in upland areas, has very narrow lobes, and the plant is very much branched.

4 **Lycopodium inundatum** is a Clubmoss (*see* p. 56), found growing on wet, sandy or peaty soil which is covered with shallow water in winter. It is commonest on wet heaths in southern England, but it also grows in a few places in Scotland and Ireland. The creeping, wiry stems are densely clothed with pale, yellowish-green, narrow and sharply pointed leaves, which are turned upwards and backwards at the tips. They are anchored firmly to the ground by numerous, tough, branching roots growing along their whole length, and are rather short-lived. Only the tips of the shoots usually survive the winter and grow on to form new plants in the spring. The cones are rather like those of *Selaginella selaginoides* (p. 57), but only one kind of very small and dust-like spore is produced.

5 **Chara hispida** (Stonewort) is the largest species of *Chara* found in Britain, and it is common throughout the country. It has long, robust stems with whorls of from nine to eleven branchlets at intervals. The sections of stem between successive whorls are spirally twisted. Reproductive structures (5A) are of two kinds. The male bodies are pale orange spheres; the female ones look rather like tiny pineapples, and are green when young, but become reddish brown when ripe. These female structures are very similar in all species of Charales, and have a quite distinctive appearance; they are known as fossils from Coal Measure times onwards.

6 **Chara aspera**, which is found throughout Britain, is a more slender and less robust plant than *C. hispida*. There are eight or nine branchlets in each whorl, and the sections of stem between them are slender and usually more than twice as long as the branchlets. The small, orange, male reproductive structures and the female ones which become black when ripe grow on separate plants. There are fifteen other species of *Chara* known in Britain, but most of them are quite rare.

1 CEPHALOZIA BICUSPIDATA   2 CALYPOGEIA FISSA   3 RICCARDIA PINGUIS
4 LYCOPODIUM INUNDATUM   5 CHARA HISPIDA   6 CHARA ASPERA

1 **Osmunda regalis** (Royal Fern) is the largest British fern. It used to grow in wet places throughout the country, but now, owing to the effects of drainage and the depredations of collectors, it is common only in the west and north. Under favourable conditions, the branching rootstock, clothed in a tangled covering of dead leaf bases and much branched black roots, forms a globular mass two or three times as large as a football. The fronds, which may be taller than a man (only part of a frond is shown here), have stout, yellowish-green stalks with narrow wings which broaden at the darker-coloured base into distinct flaps. The blades, which are about two-thirds the length of the fronds, have seven to nine pairs of main branches, each with about a dozen pairs of narrow, oblong leaflets. The spore cases are pear-shaped and reddish brown when ripe (1A), and they develop on specially modified fronds growing from the centre of the tuft. These have two or three pairs of branches with ordinary leaflets and ten or more pairs on which the spore cases develop in clusters. Both kinds of frond are clothed with a dense covering of woolly hairs when young, which falls off as they unfold, leaving them quite smooth.

2 **Dryopteris cristata** is a rare fern of wet heaths and marshes, now restricted to East Anglia and single localities in Kent, Surrey, and Renfrewshire. The other places in which it used to grow have been drained. The fronds grow in small tufts from creeping underground stems. The outer fronds are spreading, and the inner ones, which have clusters of spore cases protected by kidney-shaped scales (indusia) on their under surfaces, stand erect. The stalk of each frond, which is about one-third as long as the blade, is clothed with pale-brown, oval scales, especially at the base. The blade has ten to twenty pairs of main lobes, which are further divided, but not usually quite to the midrib. The largest lobe is at about the middle of the blade of the frond, and has five to ten pairs of smaller lobes. The margins of all the lobes are toothed, with short, stiff, bristle-like hairs at the tips. Other species of *Dryopteris* are described on pp. 184 and 186.

3 **Equisetum fluviatile** (Horsetail) grows at the margins of lakes and rivers and in marshes and swamps throughout Britain. Hollow stems grow erect from tough, smooth, purplish-black, creeping stems buried in the mud and anchored by tufts of numerous hair-like roots. They usually develop a few whorls of short branches, but sometimes they are quite unbranched. Some stems have short, thick, oval cones at their tips, which are black and green when young, as shown here, but become brown when the spores are ripe. Stems without cones taper gradually to a point. The short sheaths (3A) are closely pressed to the stem and have ten to thirty short, slender, sharply-pointed, black teeth. The stems are smooth to the touch, but have very fine grooves corresponding to the teeth of the sheaths. The central hollow cavity, which is larger than in any other species of *Equisetum*, may take up as much as four-fifths of the diameter of the stem. There is a note on *Equisetum* on p. 190.

4 **Equisetum palustre** is widespread in Britain in marshes and wet places. The slender, tough, black, underground stems descend to considerable depths in the mud and produce numerous oval tubers, each about the size of a haricot bean. The hollow upright stems usually have a few whorls of short branches, as shown here, but sometimes they are quite unbranched. Some of them develop a blunt-tipped cone (4A) on a short stalk at the tip, which is black when young, but becomes brown when ripe. The rather short sheaths on the stem have four to eight short, pointed, dark-brown or black teeth, each with a narrow white margin and a rib down the centre. The stem has grooves corresponding to the teeth of the sheaths, and inside is a ring of hollow cavities alternating with the grooves and a small central cavity, all the same size. (*See also* p. 190.)

Equisetum hyemale (Dutch Rush) is a rather uncommon plant of shady streamsides, which is becoming rarer. The unbranched stems remain green throughout the winter and have ten to thirty grooves, the ridges between them being very rough to the touch, like a medium grade of glasspaper. Because of this roughness the plant was used for cleaning cooking pots before the days of steel wool and nylon pan-scourers.

1 Osmunda Regalis   2 Dryopteris Cristata
3 Equisetum Fluviatile   4 Equisetum Palustre

FRONDS — HALF LIFE SIZE   DETAILS — MAGNIFIED

1 **Dryopteris sp. (prothallus).** The spores of ferns germinate readily on moist soil anywhere and form small, bright green, flat, leaf-like structures resembling plants of a liverwort such as *Pellia epiphylla* (p. 101); they are, however, thinner and considerably more delicate. This stage in the life cycle, which is called a prothallus and is heart-shaped and does not branch, is comparatively short-lived. Reproductive structures are produced on the underside, which eventually give rise to a tiny fern plantlet (1A) that grows on the prothallus at first, but soon develops roots and becomes independent. The plants shown here are the prothallus stage of a species of *Dryopteris* (p. 187).

2 **Riccia fluitans** is a liverwort which grows in floating masses in ponds and ditches in which the water contains moderate amounts of dissolved mineral matter. The individual fronds (2A) are narrow and repeatedly forked, and contain air chambers that give them a translucent appearance. Another form of the plant, which is less common, grows on mud, to which it is attached by tiny white hair-like structures (rhizoids). The fronds are about twice as broad and have a shallow channel along the middle of the upper surface.

Riccia glauca ('Crystalwort') grows on moist soil, especially in fallow fields, from autumn until the spring. It forms rosettes about the size of a sixpence, of shallowly forked, bluish-green fronds. *R. sorocarpa* is similar, but the fronds have a deep, narrow channel along the upper surface.

Ricciocarpus natans has free-floating rosettes, similar in size and shape to Duckweed (*Lemna minor*). The shallowly-forked frond contains air chambers, and long, thin, violet scales hang down from its lower surface into the water. This liverwort is an uncommon plant of ponds and ditches, and is occasionally found growing on mud.

3 **Azolla filiculoides** is the only free-floating fern to be found in Britain. It is a native of the warmer parts of America, but has become naturalised in ponds and ditches in scattered localities in southern England and elsewhere in Europe. The yellowish-green fronds (3A), which are frequently tinged with red, especially in the winter, consist of branched stems with two rows of leaves; each leaf has a thin, submerged lobe bearing the spore-producing structures, and an upper floating lobe which overlaps neatly the leaf in front. Numerous roots hang down from the underside of the stem.

There is a cavity in each of the upper leaf lobes which invariably contains threads of the blue-green alga *Anaboena*.

4 **Pilularia globulifera** ('Pillwort') is an aquatic fern found only in western Europe. It grows on mud at the margins of ponds and lakes in all parts of Britain, but it is not very common and is easily overlooked. The thin, creeping, rather sparingly-branched stem bears roots and narrow, cylindrical, rather stem-like leaves. It might easily be mistaken for a grass, sedge, or rush, but can be distinguished by the fact that the young leaves are spirally coiled at the tip, and by the presence in mature specimens of small spherical spore-producing sporocarps at the bases of the leaves. The sporocarps are covered with a dense coat of hair and are yellowish green when young, but turn light brown, and finally brownish black, with age.

5 **Fontinalis antipyretica** ('Willow Moss') is easily the largest moss that grows in Britain, for it sometimes forms dense, pendant bunches as long as a man's arm. It grows attached to stones and wood in slow-flowing rivers and streams, and also in lakes and ponds. Even small plants may be recognized by the characteristically keeled, dark-green leaves (5A). The specific name (meaning 'against fire') is said to refer to a former use of the plant as a non-inflammable insulating material in the walls of houses in Lapland.

6 **Isoetes lacustris** ('Quillwort') is related to the club-mosses, *Lycopodium* and *Selaginella* (p. 57). It grows, often at considerable depths, in lakes and tarns in which the water is practically free from dissolved mineral matter. It has a dense tuft of leaves, which may be three or four times as long as in the plant shown here, and abundant, thick white roots. The spore-producing structures are embedded in the bases of the leaves. *I. lacustris* may be distinguished from certain rather similar looking aquatic flowering plants, such as *Lobelia dortmanna*, which sometimes grow with it by the four longitudinal tubular air spaces with cross partitions which are present in each leaf. *I. echinospora* is similar but smaller and has paler green leaves, which are soft and flexible, rather than stiff and brittle. It is less common, and grows in lakes and pools with peaty bottoms.

Isoetes hystrix is a rare plant found in the Lizard district of Cornwall. It grows buried in damp ground, with only the tips of the leaves showing, in places which are under water in winter, but dry in summer.

1 DRYOPTERIS *sp.* (PROTHALLUS)

2 RICCIA FLUITANS 3 AZOLLA FILICULOIDES 4 PILULARIA GLOBULIFERA

5 FONTINALIS ANTIPYRETICA 6 ISOETES LACUSTRIS

The plants on this page are mosses except Nos. 4 and 7. Their branching stems are covered with leaves which have a midrib or 'nerve', and their spore capsules open by means of a small lid at the top which becomes detached when ripe. There is a note about liverworts at the head of the next page.

1 **Leskea polycarpa** usually grows on the bases of trees or on wood, or occasionally on the ground, in places liable to flooding by lowland streams where the waters contain moderate amounts of dissolved substances. It has creeping stems with numerous short upright branches clothed with crowded tiny leaves. The spore capsules are long and narrow, and are held upright on reddish-brown stalks which grow directly from the creeping stems (1A). The leaves are very small and taper to not very sharp points; the nerve does not extend right to the tip.

2 **Amblystegium serpens** is found in similar places to *Leskea polycarpa* and has a similar mode of growth, but the leaves are even smaller. It grows on wood, sometimes on rotting tree stumps, on rock, and on the ground in moist woodland. It forms low, soft patches, consisting of crowded upright branches growing from creeping stems. The narrow capsules are curved and are held horizontally on red stalks growing from the upright branches. The leaves taper to fine points; the nerve is rather faint and does not extend beyond mid leaf.

3 **Cinclidotus mucronatus** grows on rocks and on tree stumps by rivers and slow streams. It forms dark-green, short, erect tufts of stems, which are blackish brown at the base. The lower leaves are small and clasp the stem, while those towards the top are larger and more spreading (see enlarged detail 3A). They have thickened borders. The well-developed nerve projects beyond the leaf tip as a short point. The leaves become spirally twisted when dry.

Cinclidotus fontinaloides grows under water in streams and rivers, attached to submerged stones and wood. It has some resemblance to *Fontinalis antipyretica* (p. 97) for it is a larger plant than *C. mucronatus*, with longer, more branched stems; but the structure of the leaves is very similar.

4 **Trichocolea tomentella** is a liverwort which grows in very wet places in the shade. It is found in alder thickets on swampy ground by streams, or sometimes on wet, shaded rocks; but it is nowhere very common. The moderately robust stems are very regularly branched two or three times; they are densely clothed with leaves, which are deeply divided into many thread-like segments. In amongst the leaves are very numerous, even finer threads (paraphyllia).

5 **Brachythecium rivulare** is to be found in mountain springs, by waterfalls, and in and by streams generally; occasionally it grows on very wet ground in woodlands. It is a bright golden-green colour, and has creeping stems with numerous, crowded, upright branches. The leaves are tinged with orange at the base where they run down the stem and are pointed at the tip, although not finely so; the margins are finely toothed. There are several deep folds running parallel to the nerve, which extends for rather more than half the length of the leaf. Other species of *Brachythecium* are shown on p. 39.

6 **Eurhynchium riparioides** forms extensive patches on stone and wood in rapidly-flowing streams and rivers. The main stems are long and rather sparingly branched, becoming bare of leaves in their lower parts. The leaves are oval with narrow bases, acute tips, and closely and rather strongly toothed margins. The nerve extends for about three-quarters of the length of the leaf. Other species of *Eurhynchium* are shown on p. 43.

7 **Marchantia polymorpha** is a large liverwort found in marshy places and on the banks of streams, and also very commonly in gardens and greenhouses. It has a well-marked midrib, and a series of air chambers just beneath the surface forms a pattern of hexagonal markings. There is a tiny barrel-shaped pore in the centre of each marking, but these can be seen only with the aid of a strong lens. The male reproductive bodies are produced on the slightly-hollowed upper surface of stalked structures, looking rather like small toadstools, which are shown on the left of the illustration. They are dispersed by raindrops which bounce off and may be caught between the 'ribs' of the rather parasol-like female reproductive structures, shown on the right. Spore capsules eventually develop underneath the 'ribs'. Another kind of reproductive body is shown in 7A. Small green granules (gemmae) are produced at the bottom of cups with wide openings and toothed margins; these are widely scattered by rain splash and grow into new plants.

8 **Leptodictyum riparium** grows on stone, wood, or soil on the banks of streams and ponds, or actually submerged in the water. The stems are extensively but loosely branched, and the finely-pointed leaves spread rather widely. The nerves extend for three-quarters of the length of the leaves, which have untoothed margins. The curved capsules have slender, orange-red stalks.

1 LESKEA POLYCARPA   2 AMBLYSTEGIUM SERPENS   3 CINCLIDOTUS MUCRONATUS
4 TRICHOCOLEA TOMENTELLA   5 BRACHYTHECIUM RIVULARE
6 EURHYNCHIUM RIPARIOIDES   7 MARCHANTIA POLYMORPHA   8 LEPTODICTYUM RIPARIUM

STREAMS AND STREAMSIDES:
LIVERWORTS

The majority of liverworts have stems with three rows of leaves, those in one row (the 'underleaves') being smaller than the others, or in some cases missing altogether. The leaves are of various shapes, often being lobed and divided, and they have no midrib or 'nerve'. In the species illustrated here, the whole plant looks rather like a single flat, lobed leaf, which is anchored to the ground by a large number of long, fine, white hairs (rhizoids) on the underside (*see* 6A). The plants are branched in a regular forking manner, and the growing points are at the bottom of shallow notches at the tips of the branches. All liverworts produce capsules which, unlike those of mosses, are short-lived. They split open in a very regular manner in order to release the spores, the wall forming four turned-back flaps or 'valves' (*see* 1A).

1 **Pellia epiphylla** grows on rich soil in moist shady places, especially on the banks of streams. The regularly-forked branches have broad, rather thin midribs, and are a shining dark green, and somewhat translucent. Male reproductive structures develop in minute pits in the upper surface, and female ones under small flaps (involucres) above the midrib. In the early spring, the very dark greenish-brown or almost black capsules are produced abundantly at the tips of long, thin, semi-transparent stalks. When they split open, the spores are seen to be entangled in a tuft of pale, greenish-brown hairs (elaters); small variations in the amount of moisture in the air cause these to twist in a jerky manner, thus gradually dispersing the spores.

2 **Pellia fabbroniana** is found growing on moist, shaded, limestone rocks or on wet chalky soil. The branches are narrower than those of *P. epiphylla* which, also, never grows on chalk or limestone. In the autumn and winter, tufts of small forked branchlets are produced at the tips of the branches (2A). They are very brittle and can break off and grow into new plants.

3 **Conocephalum conicum** grows on wet, shaded boulders of limestone and other rock and on walls and stonework in rivers and streams. Normally it has broad, dark-green, thick and fleshy branches, but in poor conditions and inadequate light elongated branches only half the usual width and both thinner and paler may be found. This often happens when attempts are made to cultivate this attractive plant. It has a pleasant, fragrant smell when bruised. A series of air chambers just beneath the upper surface is visible as a pattern of hexagonal markings, with a conspicuous circular opening in the centre of each. Male reproductive structures are produced in small, slightly-raised, purplish-green cushions near the tips of the branches, and female ones on the underside of a cone-shaped hood borne on a tall stalk.

4 **Blasia pusilla** is to be found on the banks of streams and on wet, clayey or gravelly soil generally. It is pale green, and has hollow chambers on the underside containing threads of the blue-green alga *Nostoc*, which appear as dark dots on the upper surface (4A). There are also small, toothed underleaves on the undersurface. The capsules are similar to those of *Pellia epiphylla*, but oval rather than spherical. Two other kinds of reproductive bodies (gemmae) that are sometimes found are tiny star-shaped scales just behind the branch tips, and small granules produced in flask-shaped structures with long necks.

5 **Preissia quadrata** grows on rocks in streams or on damp soil in limestone areas; it is quite common in upland areas in the north and west of Britain. Like *Conocephalum conicum*, it has a pattern of hexagonal markings with openings in the centre, but it is a much smaller plant and is pale rather than dark green. Male reproductive structures are produced on discs with short stalks, and female ones on the undersides of four-lobed caps with much longer stalks, as shown in the illustration.

Reboulia hemisphaerica is found on soil amongst rocks and on sheltered hedge-banks. It is similar to *Preissia quadrata*, but it has a much larger number of air chambers in more than one layer, so that when cut across it appears honeycombed with air spaces. Also the stalked cap with the female reproductive structures usually has more lobes, even as many as seven.

6 **Lunularia cruciata** is common in gardens and greenhouses and is not found far away from them. It was probably introduced accidentally into this country from the Mediterranean region long ago with imported cultivated plants. The air spaces with their openings show only rather indistinctly on the surface. Green, granular reproductive bodies (gemmae) are produced in crescent-shaped, incomplete cups (6A). Other forms of reproductive structures are very rare, though female ones are occasionally found on the underside of a four-lobed cross-shaped structure on a long stalk.

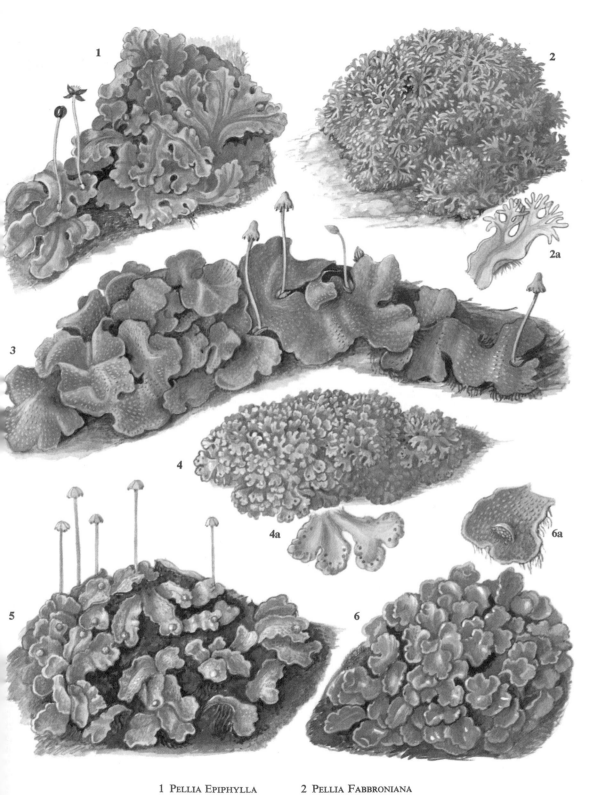

1 Pellia Epiphylla 2 Pellia Fabbroniana
3 Concephalum Conicum 4 Blasia Pusilla
5 Preissia Quadrata 6 Lunularia Cruciata

Woodlands provide many situations in which flowerless plants may grow — on the trees themselves as well as on the ground amongst them. The trees have a profound effect on the habitat as a whole; for instance, on the soil (*see* p. 184). They also provide shelter from sun, wind, and rain. The amount of shade varies from slight in open birch woods to very considerable in beech woods and beneath conifer thickets (*see* p. 158). Although the trees provide shelter from the direct effects of rain, the atmosphere in woodlands is more humid than in the open (*see* p. 174). Many fungi are parasitic upon the trunks and branches of living trees; others live on dead trunks or stumps. Other fungi which live in the ground may enter into a close relationship with the roots of trees. The degree to which a fungus is restricted to particular kinds of tree varies. Those which are specific (that is, restricted to one tree species only) are grouped as far as possible under the appropriate headings, while those which are associated with a variety of trees are described under Mixed Woodlands. For a note on Conifer Woods, *see* p. 104.

1 **Hygrophoropsis aurantiaca** ('False Chanterelle') is found in conifer woods and on heaths in late summer and autumn. The funnel-shaped cap is bright orange, and so is the stalk, but the gills, which run down the stalk, are a darker orange red. The spores are white. This fungus is harmless if eaten, but the taste is not pleasant. It has been called *Cantharellus* or *Clitocybe aurantiaca*, but it lacks the distinctive apricot smell of *Cantharellus cibarius* (p. 123), and its colour is orange rather than yellow, and the pale-orange flesh is soft, resembling cotton wool in texture. It may grow in similar places to *Clitocybe infundibuliformis*, but the cap of this species is pale brown with a distinctly pinkish tint, the stalk is paler and slightly swollen and hairy at the base, and the gills as well as the spores are white.

2 **Hygrophorus hypothejus** grows in pine woods in late autumn. The brownish-olive cap is slightly depressed in the centre and is slimy to the touch. The stalk is a paler olive brown and is also slimy, except at the top. The yellow gills, which have a waxy texture, run down the stalk and are rather distantly spaced. The spores are white. Other species of *Hygrophorus* are described on p. 32.

3 **Nolanea cetratus** grows amongst mosses in conifer woods in autumn. The cone-shaped cap is wider than it is tall, like a 'Chinese Hat'. It is yellowish brown and, when moist, is grooved from the margin half way to the centre; when dry, it is smooth, shining, and paler. The stalk is tall and slender and pale yellowish brown. The gills are pale yellow when young, but become tinged pink as the salmon-coloured spores develop.

4 **Collybia maculata** grows in groups in summer and autumn, usually associated with conifers. The whole plant is white when young, but soon becomes covered with brownish-red spots. The cap has a strongly incurved margin, and the stalk is tough (cartilagenous). The gills, which are predominantly white, although spotted with brownish-red blotches, are narrow and crowded. The spores are white. Other species of *Collybia* are described on pp. 123, 127, and 133.

5 **Agaricus silvaticus** ('Pine-Wood Mushroom') is found growing on conifer-needle litter in the autumn. The cap is covered with reddish-brown scales; the stalk is white, with a well-developed ring. The gills are pink when young, but become brown as the chocolate-coloured spores develop. The flesh is white when young and turns pink when cut; in old specimens it has a brownish tinge. *A. haemorrhoidarius*, which is similar but much rarer, has flesh which turns deep red on exposure to air. Both species are edible, but not much liked on account of the colour of the flesh.

6 **Agaricus silvicola** ('Wood Mushroom') grows in various kinds of wood in late summer and autumn. The cap and stalk are white when young, but develop a yellowish tinge with age; they also stain yellow when bruised or cut. There is a large ring consisting of two layers that hang loosely downwards; the lower layer is frequently split into tooth-like lobes. The gills are greyish white with a faint flesh tint when young, but become brown as the chocolate-coloured spores develop. The flesh is a pale cream colour and smells of aniseed; it is good to eat.

7 **Cystoderma amianthinum** grows amongst mosses in conifer woods in autumn. The cap is yellow, tinged with brown, and is toothed at the margin. The stalk is yellow and has a distinct but inconspicuous ring. Both the cap and the stalk below the ring are covered with minute granules. The skin turns a rusty brown when treated with a drop of potassium hydroxide solution. The gills and spores are white.

1 Hygrophoropsis Aurantiaca   2 Hygrophorus Hypothejus

3 Nolanea Cetratus   4 Collybia Maculatus

5 Agaricus Silvaticus   6 Agaricus Silvicola   7 Cystoderma Amianthinum

The only conifer woods in Britain which are even partly natural are some open pine woods in the highlands of Scotland, which support a rich growth of flowerless plants. The only other two native conifers, yew and juniper, can scarcely be said to form woodlands. Few plants are associated with yew; and while juniper is still common in Scotland, in southern England it seems to be dying out. This may be because it is attacked by the minute fungus *Lophodermium juniperinum*, the fruit bodies of which can be seen as minute black dots on the dead leaves of infected bushes. Many conifer woods throughout the British Isles have been planted with trees not native to Britain. In these plantations, plants other than fungi are discouraged by the low light intensity due to close planting and by the slowly decaying and highly acid conifer-needle litter. Conifer bark is acid in reaction, poor in minerals, and flakes off very readily, and thus rarely supports a very luxurious growth of mosses, liverworts, and lichens.

1 **Lactarius deliciosus** ('Saffron Milk Cap'). The whole plant, which appears in late summer and autumn, is orange coloured and develops a greenish tinge with age; it also stains green when bruised. The cap is usually slightly hollow in the centre and incurved at the margin; it is marked with indistinct alternating lighter and darker zones. The flesh is brittle and, when it is broken, exudes a tasteless saffron-coloured milk which rapidly turns carrot coloured and acquires a faintly acrid flavour. This fungus is edible, but most people would agree that *deliciosus* is a misnomer; it is said to taste better if washed very thoroughly before cooking. It has the effect of turning the urine red. It is possible that Linnaeus, who named it, may have confused it with the south European species *Lactarius sanguifluus*, which tastes better and does not turn the urine red.

2 **Lactarius hepaticus** ('Liver Milk Cap') grows in pine woods in late autumn. The cap is dark brown, the gills pink, and the stalk cinnamon coloured flushed with orange. The white milk turns yellow when exposed on the gills or on the finger, though on a knife blade or on a glass microscope slide this colour change does not occur.

3 **Lactarius camphoratus** ('Curry Milk Cap') is found in late summer and autumn, nearly always associated with conifers. The cap and stalk are reddish brown, and the gills are pink. The milk is rather watery and has the appearance, consistency, and tastelessness of whey. Very old or dried specimens develop a characteristic odour reminiscent of an inferior brand of curry powder.

4 **Lactarius rufus** ('Red Milk Cap') is found in pine woods in late summer and autumn. The cap is reddish brown, and the stalk is paler, especially at the base. The cream-coloured gills are rather crowded and may become stained with reddish-yellow blotches as they age. The milk is white and after about a minute tastes very strongly peppery. A further distinction from *L. camphoratus* is that the surface of the cap is finely granular in texture.

Other species of *Lactarius* are shown on pp. 113, 121, 127, and 131.

5 **Mycena epipterygia** ('Yellow Stalk') grows in woods and on heaths in late summer and autumn, and is usually associated with conifers. The yellowish olive-green cap has an elastic, slimy skin which is easily removed. The yellow stalk is also slimy. The gills are white with a pinkish tinge, and their margins may be easily detached as slimy threads. The spores are the colour of cartridge paper.

6 **Mycena rubromarginata** ('Pink Bonnet') is found growing in groups in pine woods in the autumn. The cap is brown when wet, but becomes pinkish grey on drying; the gills are pale greenish brown edged with pink. The flesh smells of ammonia. The spores are creamy white.

7 **Mycena lactea** ('Snow Bonnet') grows in groups under pines in the autumn. The whole toadstool is white and is quite odourless. The gills are crowded, and the spores are white.

8 **Mycena cinerella** ('Mole Cap') is found in the autumn, usually growing in groups and associated with conifers. The cap is grey, and the gills are greyish white with pink edges. The flesh smells of new meal. The spores are creamy white.

Other woodlands species of *Mycena* are shown on pp. 127, 133, and 145, and grassland species on p. 33.

1 Lactarius Deliciosus

2 Lactarius Hepaticus   3 Lactarius Camphoratus   4 Lactarius Rufus

5 Mycena Epipterygia   6 Mycena Rubromarginata   7 Mycena Lactea   8 Mycena Cinerella

1 **Tricholomopsis rutilans** grows on or near pine stumps in late summer and autumn. The whole plant is yellow, but the cap and stem are covered with reddish-purple downy scales, the cap densely and the stem less thickly, especially on the lower part. This fungus is always found directly on wood, or connected with it by means of branching root-like cords running from the base of the stem.

Tricholomopsis platyphylla grows on or connected with the stumps of various trees other than conifers in a similar way to *T. rutilans*. The cap is smoky grey and is marked with thin dark radial streaks. The white gills are very broad and distantly spaced from one another.

2 **Russula sardonia** is a dark purplish-violet colour, which becomes paler with age. The stalk is flushed with the same tint, and the gills are distinctly tinged yellow. The white or pale lemon-coloured flesh has a very acrid taste and turns red when treated with a drop of ammonia (50% ammonium hydroxide solution); the colour may take ten minutes or so to develop. (See general note on *Russula*, p. 136.)

Russula queletii has a wine-red or purplish-violet cap, and the stalk is flushed with the same colour. The gills are ivory white when young, but become cream or creamy grey with age. The white flesh is usually faintly tinged green and tastes very acrid.

3 **Russula coerulea** is a dark purplish-violet colour. The cap has a small darker-coloured boss (umbo) at the centre. The gills are cream coloured when young, developing a brownish tinge with age, and the spores are yellow.

4 **Russula nauseosa** is reddish-brown with a distinct purplish pink tint when young, which fades to pale yellowish brown with age. The gills, cream at first, become deep yellow or almost orange as the egg-yolk-coloured spores develop.

5 **Russula puellaris** has a purplish-pink cap, which is darker in the centre. The whole plant becomes a characteristic yellow colour, usually described as 'wax yellow', as it ages.
R. versicolor, which is associated with birch, is very

similar to *R. puellaris*. However, the dark centre of the cap is usually tinged with green, and the spores are a darker yellowish white.

6 **Clitocybe langei** is found in conifer woods and on heaths, frequently growing under bracken, in late autumn and early winter. The whole plant is brownish grey — in fact, a 'mousey' colour. The cap, which is usually slightly hollowed in the centre, has numerous fine radial grooves at the margin when it is wet; on drying, these disappear, and the colour changes to a greyish white. The gills are paler than the cap, and the smooth, elastic stalk is paler still. The flesh tastes and smells of meal.

7 **Clitocybe clavipes** ('Club Foot') grows in conifer woods and also under beech trees in late autumn and early winter. The greyish-brown cap of younger specimens has a slight boss (umbo) in the centre, which in older specimens becomes a shallow depression. The pale primrose-yellow gills run deeply down the stalk, which tapers upwards from a swollen base. The enlarged spongy base resembles that of *Collybia butyracea* (p. 133), which also has a greyish-brown cap and is found in similar habitats. However, the latter has a brown umbo which persists, gills which do not run down the stem, and a tough stalk.

8 **Clitocybe nebularis** is found in various kinds of woodland, but especially amongst conifers, in late summer and autumn. The whole plant is a pale grey colour. The cap has a broad boss (umbo) in the centre when young, which with age becomes flattened and eventually replaced by a slight depression. The gills are paler than the cap, and may be pure white in young specimens; they are crowded and run slightly down the stalk, which tapers upwards from the base. The white flesh has a faintly bitter-sweet taste and a rather sickly smell. It is edible but should not be eaten in quantity.

9 **Clitocybe flaccida** grows in clusters or sometimes in rings in conifer woods in late autumn and early winter. The cap is reddish brown and shaped like a shallow funnel. The gills are pale yellow when young and develop a reddish tint with age; they run down the stalk, which is a rather paler colour than the cap. The flesh is white when young but becomes rust brown in older specimens; it has a faintly sour smell and a rather bitter taste. It is not poisonous but is not recommended for eating.

1 TRICHOLOMOPSIS RUTILANS   2 RUSSULA SARDONIA   3 RUSSULA COERULEA
4 RUSSULA NAUSEOSA   5 RUSSULA PUELLARIS   6 CLITOCYBE LANGEI
7 CLITOCYBE CLAVIPES   8 CLITOCYBE NEBULARIS   9 CLITOCYBE FLACCIDA

There is a general note on *Boletus* at the head of p. 142.

1 **Boletus bovinus** grows in woods and on heaths under pine trees from summer to autumn. It has a slimy, pale-brown cap, tinged with red. The tubes, which run markedly down the stem, are yellow when young, but become olive brown with age; they have rather large, angular openings, which are divided into smaller pores beneath the surface. The stem tapers downwards towards the base. The flesh is pale yellow, flushed with pink, and has a strong, somewhat fruity smell.

2 **Boletus luteus** is found growing amongst grass in conifer woods. The cap is brown, tinged with purple, and is very slimy to the touch. The yellow tubes are covered by a thin white veil when young, which remains as a ring round the stem in older specimens. The flesh is pale yellow and has a somewhat fruity smell. It is good to eat, but it is better to peel the skin from the cap and remove the tubes before cooking.

Boletus granulatus is similar to *B. luteus*, and grows in the same kinds of situation. It is, however, a paler brown colour and has no ring on the stem. When young, the tubes exude milky drops of moisture. The pale yellow flesh is good to eat.

3 **Boletus variegatus** grows in woods and on heaths on poor soils, especially under pines. The cap is somewhat slimy to the touch when wet, and is yellowish brown, sprinkled with small, soft, dark-brown scales. The tubes are yellowish grey when young, becoming brownish olive with age; they stain faintly blue when bruised. The yellowish-brown stalk is slightly swollen and tinged with red at the base. The flesh is pale yellow and becomes blotched with blue when cut; it tastes rather sour and has a faint smell of chlorine, rather like an indoor swimming bath. It is edible but of indifferent quality, and is sometimes eaten pickled in vinegar.

4 **Boletus elegans** is found only in association with larch, a tree which appears to have been first introduced from Europe in the 16th century. Both the tree and the fungus, which is found from early spring until the autumn, have become thoroughly naturalized. The slimy cap is a clear yellow, and the tubes are sulphur yellow staining dull red when bruised. The stem has a well-developed ring. The flesh has some-what the texture and colour of rhubarb, like *B. piperatus* (p. 142), and stains faintly lilac when cut. It has a pleasant smell and is good to eat.

Boletus aeruginascens is also associated with larch. The slimy cap is whitish brown, wrinkled, and covered with scattered, small, dark greenish-brown scales. The tubes have large irregular openings and are grey when young, becoming brown with age. The white flesh has a greenish tinge and turns light blue when cut.

Boletus tridentinus is unusual in that it sometimes grows under yew, a tree with which very few fungi are associated. It also grows with larch, but is not at all common. It has an orange or reddish-brown slimy cap, sprinkled with small dark-brown scales. The tubes are orange yellow and have large angular openings; they stain red when bruised. The pale yellow flesh turns salmon pink when cut.

5 **Boletus badius** grows in all kinds of woodland, but is especially associated with conifers. It is somewhat like *B. edulis* (p. 143) in appearance, the caps of both plants looking rather like sticky buns. *B. badius*, however, is chestnut brown, the yellow tubes turn green when bruised, and the stalk is not swollen. The pale yellow flesh turns pale blue when cut. It has a pleasant flavour but is not as good to eat as *B. edulis*.

6 **Gymnopilus penetrans** grows attached to sticks and chips of wood in conifer woodlands. It is yellowish orange or golden brown, old specimens becoming tinged a rusty colour or blotched with reddish-brown spots. The cap has a slight boss (umbo) in the centre and is covered with very tiny scales. When young, the gills are covered by a delicate veil or cortina, but this does not leave a ring on the stem. The spores are brown. The pale yellow flesh has a very bitter taste and a distinct smell, said to resemble that of birch leaves.

Gymnopilus junonius is similar to *G. penetrans*, but is at least twice as large and has no umbo. There is a well-developed ring on the stalk. It grows in tufts at the base of conifers and also other types of tree.

1 Boletus Bovinus   2 Boletus Variegatus   3 Boletus Luteus   4 Boletus Elegans
5 Boletus Badius   6 Gymnopilus Penetrans

Bracket fungi obtain their nutriment either from food substances stored in the sap wood of the trees on which they grow, or by digesting and absorbing the substance of the wood itself; in the latter case the heart wood will be affected as well.

Although some species are found only on dead wood, many of them attack living trees and are very destructive; in fact, in virgin forest the death of most old trees is due to bracket fungi. It is interesting to note that very long-lived trees, such as yew and the giant redwoods of California, appear to be almost completely immune from the attacks of bracket fungi, and in the rare cases where a fungus does succeed in establishing itself, it is able to grow only extremely slowly.

Other species of bracket fungus are shown on pp. 117, 125, 129, and 157.

1 **Fomes annosus** is sometimes found on the trunks of all kinds of conifers, and sometimes on other trees and on felled timber. The upper surface has a thin, hard, resinous crust; when young it is pale brown and thinly velvety, but in older specimens it may be shiny and is dark brown, usually with a reddish tint and sometimes with indistinct alternating lighter and darker zones. The tubes are yellowish white; the hard, woody flesh is white or pale buff; and the spores are white. The fungus is perennial and causes a serious heart rot in conifers. In Britain it attacks spruce especially.

2 **Trametes abietina** grows on stumps and dead branches of conifers, often forming tiered clusters of brackets with wavy margins. The softly velvety upper surface is whitish or ashy grey, sometimes with violet tints near the edges and often indistinctly zoned. The tubes have angular openings and are pale brown when young, becoming darker with age, but always strongly tinged violet. This species is said to be especially common where trees have been destroyed by fire, but wherever there are dead trunks of conifers it is as common as *Trametes versicolor* is on other kinds of timber (p. 116).

3 **Phaeolus schweinitzii** grows at the bases of conifers. Very often it has a thick stalk which is centrally placed, but this may be joined to the edge of the bracket, or may be absent altogether. The upper surface is orange to rusty brown and is thickly covered with rather stiff woolly hairs. The stalk is similar. The tubes are yellow with a greenish tint and become darker when bruised or dried. The yellowish to reddish brown flesh is spongy and rather watery, and the spores are white. This fungus causes a serious rot in the lower part of the trunk of any tree on which it grows, causing the wood to break up into reddish-brown cubical blocks. In Britain it attacks pine especially.

4 **Leptoporus caesius** is found on stumps and dead wood of various conifers. The upper surface is white or whitish grey, usually strongly tinged with blue or with a blue zone at the margin, and covered with soft short hairs. The tubes are white or ashy grey, tinged with blue, or staining pale blue when bruised, and have unequal, somewhat angular openings. The flesh is soft and rather watery with a mild taste and a faintly sweet smell. The spores are pale ashy blue.

5 **Leptoporus stipticus** grows in summer and autumn on the dead wood of conifers. The white or yellowish white upper surface, sometimes tinged reddish brown at the margin, is covered with very short soft hairs. The tubes are white or cream and frequently have irregular openings. The flesh has a texture rather like that of soft cheese, and is extremely bitter and astringent to the taste. The spores are white. During active growth, copious drops of water are exuded from the tube surface.

Poria versipora grows in patches which lie quite flat and firmly attached on the bark of fallen branches of all kinds of trees, with the layer of tubes uppermost. It is irregular in outline, with a very uneven surface, is usually pure white at the margins, and cream coloured towards the centre. The tubes have wide openings with toothed margins, and the spores are white. *P. vaillantii* is similar, but softer and looser in texture, with a cottony appearance at the margins. It is more easily detachable from the wood on which it grows. It causes decay of timber in wet situations. There are perhaps as many as fifty species of *Poria* to be found in Britain, but they can be identified correctly only by careful examination of their microscopic structure.

6 **Rhizina undulata** grows flat on the ground, associated with dead twigs, bark fragments, wood chips, and charcoal, in conifer woods in summer and autumn. It forms broad, flat cushions which are frequently irregularly lobed and turned under at the margins. The upper spore-producing surface is very dark brown or almost black, although somewhat paler at the edges. The reddish-brown flesh is tough and fibrous. On the under surface there are numerous root-like structures (rhizines). This fungus, which is especially common after forest fires, appears to be associated with burnt ground in conifer woods. It is regarded as a serious pest in plantations because it is thought to attack seedling pines.

1 FOMES ANNOSUS   2 TRAMETES ABIETINA
3 PHAEOLUS SCHWEINITZII
4 LEPTOPORUS CAESIUS   5 LEPTOPORUS STIPTICUS   6 RHIZINA UNDULATA

Birch, together with pine, was the first tree to invade Britain when the glaciers retreated at the end of the ice age. It is still able to hold its own and colonizes cleared areas rapidly, especially on poor soils. It is not a long-lived tree and casts only a light shade at all stages of its growth, so the ground flora in birch woods is often rich. Birch bark, which is very smooth in young trees and peels off readily, is, like conifer bark, acid in reaction and poor in minerals. Consequently it rarely supports a very luxurious growth of mosses, liverworts, and lichens.

1 **Amanita muscaria** ('Fly Agaric') is found in late summer and autumn, sometimes in pine woods as well as birch woods, and usually on poor soils. The scarlet cap and the white stalk and gills are at first completely enclosed in a white membrane, the torn remains of which may be seen as a volva at the base of the stalk and as patches adhering to the surface of the cap. In old specimens, especially in wet weather, the patches are washed or rubbed off, and the colour fades to a paler orange red. There is a well-developed ring on the stalk. This is the most frequently illustrated of all fungi. The earliest known picture of it is in a famous fresco in a ruined church at Plaincourault in France, dating from 1291. This is not, however, the earliest known illustration of a fungus. In a 1st-century fresco at Herculaneum there is a picture of *Lactarius deliciosus* (p. 105). *A. muscaria* is poisonous, though unlikely to be fatal; Albertus Magnus in the 13th century recommended it, broken up in milk, as a fly killer. It has been used as a drug to produce hallucinations at various times in different parts of the world.

2 **Amanita fulva** ('Tawny Grisette') grows in woods, usually associated with birch and especially on peaty soil, from late spring until autumn. The cap is orange brown, the stalk a similar but much paler colour, and the gills are white. There is a deep, conspicuous cup (volva) at the base of the stalk. This fungus has sometimes been called *Amanitopsis fulva* because there is no ring on the stalk. Though it is not poisonous, the flesh is thin and fragile and not of great value in the kitchen.
A. vaginata ('Grisette') is similar to *A. fulva*, with a well-developed volva but no ring, but the whole plant is a pale or ashy grey colour. It grows in woods and on heaths, frequently with beech trees.

3 **Lactarius turpis**, which grows under birch trees on damp peaty soil, is dark olive brown and somewhat slimy to the touch. The cap is sometimes almost black except for the margins, which are tinged yellow, in-rolled, and covered with woolly hairs. It changes to a deep purple when treated with a drop of ammonia (50% ammonium hydroxide solution). The surface of the stalk is marked with numerous small pits. The crowded gills are a pale straw colour, becoming brown when bruised. When broken, the white flesh

exudes a white milk (latex) which tastes very peppery. All the *Lactarius* species described on this page appear in late summer and autumn. Their brittle flesh exudes a milky latex when broken, and the spores are white or cream coloured.

4 **Lactarius vietus** is found in damp woods, sometimes associated with pine as well as birch. The cap is pale grey tinted with flesh pink and lilac and is softly downy at the margins. The gills, white when young, become a yellowish flesh colour with age and stain olive brown if bruised or broken. The white peppery-tasting milk becomes a grey colour on drying, after perhaps 20 minutes or more.

5 **Lactarius tabidus** usually grows amongst Sphagnum (p. 87) in damp woods, especially under birch. The cap is pale brown with a yellowish tinge, and the upper surface is creased around a central boss (umbo). It is rather like *L. subdulcis* (p. 131) in appearance, but the mild-tasting milk turns yellow on exposure to air for about a minute.

6 **Lactarius torminosus** ('Woolly Milk Cap') is found in woods and on heaths, usually on poor soils and associated with birch. The cap has a hollow in the centre and is sometimes almost funnel-shaped; it is strawberry pink or flesh coloured, with alternating lighter and darker zones, slightly slimy to the touch when wet, and shaggy with woolly hairs along the margin. The milk is white and tastes very peppery.

7 **Lactarius glyciosmus** grows in damp birch woods. The grey cap is tinged with lilac, and the upper surface, which is covered with very tiny scales, becomes slightly hollowed with age, although there is always a small pointed umbo in the centre. The milk is white and has a peppery after-taste. The whole plant smells strongly of coconut.
L. mammosus also smells of coconut, but is less common and is associated with conifers. It is pale brown in colour. Other species of *Lactarius* are illustrated on pp. 105, 121, 127, and 131.

1 Amanita Muscaria   2 Amanita Fulva   3 Lactarius Turpis   4 Lactarius Vietus
5 Lactarius Tabidus   6 Lactarius Torminosus   7 Lactarius Glyciosmus

There is a general note on *Russula* on page 136 and on *Boletus* on page 142.

1 **Russula claroflava** is a bright, clear yellow, though the gills become very pale yellow with age. The stalk becomes blotched with grey when bruised or cut and may be uniformly pale grey in old specimens. It is a clearer yellow colour than *R. ochroleuca* (p. 137).

2 **Russula aeruginea** is green, often grass green. The gills turn yellow with age, and the whole plant stains brown when bruised or cut. The only other common green species of *Russula* is *R. virescens* (p. 121).

3 **Russula nitida** frequently grows amongst *Sphagnum* (p. 87). It is pinkish or purplish red and is darker towards the centre; the colour fades somewhat with age. The margin is deeply grooved. Somewhat similar in colour are *R. sardonia* (p. 107) and *R. xerampelina* (p. 137).

4 **Cortinarius semisanguineus** grows in woods and on heaths, under conifers as well as amongst birch trees, in late summer and autumn. The cap is yellowish or olive brown when young, but turns reddish brown with age. The stalk is similar but paler, with brownish-olive threads on the surface, and sometimes covered with pinkish-red down at the base. The gills are a clear red colour. About 200 species of *Cortinarius* have been recorded in Britain, but few are common, and they are all difficult to identify. Young specimens may be recognized by the cobweb-like veil (cortina) covering the gills, although this often disappears completely with age, and by the rust-brown spores.

5 **Boletus testaceoscaber** is found in summer and autumn in conifer as well as birch woods. The brownish-orange cap is downy; the stalk is covered with rough black scales; and the pale greyish-brown tubes have very small openings. The white flesh turns

pink and then lilac, and after a considerable time, slate colour; it often has a bluish-green tinge at the base of the stalk. It is quite good to eat.

6 **Boletus scaber** grows in birch woods in summer and autumn. The greyish-brown or dark brown cap is slightly sticky to the touch; the tubes are pale greyish-brown, becoming blotched a darker colour when bruised; and the flesh is white. The stalk is covered with vertical rows of rough brown or black scales. Young specimens are quite good to eat.

7 **Tricholoma fulvum** is found in damp places in birch woods in the autumn. The reddish-brown cap is somewhat slimy to the touch and has a raised boss (umbo) in the centre. The similarly-coloured stalk is paler at the top; it is pointed at the base and becomes hollow with age. The gills are pale yellow when young but become blotched with brown spots in older specimens. The flesh of the stalk is yellow and of the cap, white, and it has a somewhat mealy smell and an unpleasant rancid taste.
Other species of *Tricholoma* are shown on pp. 119, 127, and 135.

8 **Paxillus involutus** grows in woodlands, usually associated with birch trees, in late summer and autumn. The cap is brown, often with an olive tinge, and is slightly slimy to the touch when moist; it has an inrolled, grooved margin, fringed with very short fine hairs. With age, a marked depression develops in the centre. The gills, which are very easily detachable from the cap, are similarly coloured and run about a third of the way down the short stalk; they are crowded, and join together to form a network at the top of the stalk. The yellow flesh becomes reddish brown when bruised or cut; it is soft and has a sour taste. It may cause illness if eaten raw, although it is said to be harmless when cooked.

1 RUSSULA CLAROFLAVA   2 RUSSULA AERUGINEA   3 RUSSULA NITIDA
4 CORTINARIUS SEMISANGUINEUS   5 BOLETUS TESTACEOSCABER   6 BOLETUS SCABER
7 TRICHOLOMA FULVUM   8 PAXILLUS INVOLUTUS

1 **Fomes fomentarius** is almost entirely restricted in Britain to birch woods in the Scottish Highlands, though there is one locality, at Knole Park in Kent, where it grows on beech. In Europe and America it is common on beech and has been found on maple, poplar, and other trees. It is perennial and causes a white rot of the timber. When young, the upper surface is pale brown to greyish brown, and thinly velvety. In older specimens it is light to very dark grey, sometimes with alternating lighter and darker zones, and it has a hard crust which, when cut, has the appearance of freshly broken resin. The tubes, when young, are pale grey, becoming buff or greyish brown with age. The spores are white. This fungus can be confused with *Ganoderma applanatum* (p. 125), but in the latter the spores are brown.

2 **Trametes hirsuta** appears on fallen branches of a wide variety of trees in summer, and may be found throughout the autumn and winter until the following spring. The upper surface is usually uniformly greyish or yellowish brown, though it is sometimes darker at the margin; it has a thick coat of short hairs, and is covered with alternating ridges and furrows. The tubes have thick walls, which make the surface uneven; they are white when young, but become a smoky yellow colour with age.

3 **Trametes versicolor** is found at all times of the year, often growing in tiered clusters, on the dead wood of a variety of trees. The upper surface is velvety and marked with alternating zones of various shades of yellow, brown, and grey. The colour is very variable, and tinges of red, green, or blue are not uncommon. 3A shows a dark form for comparison with the light form, 3. It also sometimes grows turned upwards (3B) so that the white or pale yellowish brown tubes are exposed. With a hand lens it is possible to see that they have angular openings.

Trametes gibbosa is similar to *T. hirsuta* in that the upper surface is densely covered with short hairs, but it is up to three times as large. It is greyish white, and the tubes have radially elongated openings.

Trametes cinnabarina, which has an orange-red upper surface and tubes that are the dark red colour of the mineral cinnabar, is occasionally found in this country, although it is commoner in the tropics.

4 **Piptoporus betulinus** is entirely restricted to birch trees. The upper surface is covered with a smooth, pale greyish-brown skin which is easily peeled off. The margins are inrolled and project downwards below the white tubes, as shown in the young specimens illustrated in 4. Older specimens, such as the one shown in 4A, frequently harbour the small black beetle *Cis bilamellatus*, which was accidentally introduced into Kent from Australia at the end of the 19th century, and is now widespread in south-east England, the Midlands, and south Wales. The flesh of this fungus is pure white and in fresh young specimens is firm, flexible, and moist. But it acquires a corky texture when dry and, under the name 'Polyporus', strips of it are used by entomologists for mounting small insects for display purposes. It has also been recommended for stropping razors, and is sometimes referred to in books as the 'Razor Strop Fungus'.

5 **Hymenochaete rubiginosa** grows throughout the year on the dead trunks and stumps of various trees, usually forming clusters of rather elongated narrow brackets with wavy margins. The upper surface is chestnut brown and velvety, often with alternating darker zones. The lower, spore-producing surface is rusty brown and covered with very small dark bristles. Sometimes, instead of forming a bracket, the fungus lies entirely flat on the wood, with the spore-producing surface upwards. This is especially likely to occur when it is growing on the underside of a fallen branch.

Corticium coeruleum forms shiny clear blue patches on fallen branches and twigs; in old specimens the surface becomes cracked. This fungus glows in the dark with a pale greenish-yellow light.

Coniophora arida forms very thin, powdery, greyish-yellow patches, which become dull brown with age. It is distinguished from the other related fungi by its pale brown spores.

Phlebia merismoides forms extensive orange-pink patches on stumps and fallen tree trunks. Its flesh is tough and much wrinkled with radiating vein-like ridges, and the surface is soft to the touch.

1 Fomes Fomentarius
2 Trametes Hirsuta   3 Trametes Versicolor
4 Piptoporus Betulinus   5 Hymenochaete Rubiginosa

117

1 **Amanita phalloides** ('Death Cap') is found commonly in beech woods, and also under oak trees in mixed woodlands, in late summer and autumn. It develops at first completely enclosed within a white membrane (universal veil), and when later the cap bursts through, this remains as a cup at the base, called a volva. The cap is a very pale olive green, often with a yellowish tinge, and is slightly slimy to the touch. The gills and stalk are white.

This fungus, and the related *A. virosa* (p. 131), are responsible for the great majority of the deaths that occur in Europe from fungus poisoning. The symptoms are intense abdominal pain and eventually coma and paralysis. But unfortunately no discomfort is felt until some 12 hours after eating the fungus — which John Ramsbottom suggests might even be taken as a diagnostic sign. By then, however, all ordinary methods of removing it from the stomach are quite ineffective. There are antidotes, but these are only effective if administered very soon after swallowing the poison. The only safe course is to be able to recognize *A. phalloides* with certainty and to avoid it and, if in any doubt, any fungus resembling it. It can be distinguished from white edible mushrooms by the white gills (the edible kinds have grey or pink or dark brown gills) and by the volva (not present in mushrooms). Also, the flesh of the cap turns a pinkish lilac colour when treated with a drop of concentrated sulphuric acid.

2 **Amanita citrina** ('False Death Cap') grows in beech and oak woods in summer and autumn. The cap is pale yellow and usually has a few irregular patches of the white membrane, the universal veil, adhering to it. The gills and stalk are white. The base of the stalk is swollen, and the volva, which is short and even (not large and irregularly bag-shaped as in *A. phalloides* and *A. virosa*), projects from it so as to enclose a distinct, shallow groove. The flesh smells strongly of raw potatoes and has an unpleasant taste, but is apparently not poisonous.

3 **Amanita citrina** var. **alba** is a quite common, wholly white variety, which has the same distinctively-shaped volva at the base of the stem and characteristic smell.

4 **Inocybe maculata** is found amongst beech and sometimes other trees on rich soil in the autumn. The cap is cone shaped, and the surface is radially cracked and covered with small thin scales, especially towards the centre. The stalk is a similar colour to the cap, and the gills are pale at first, but darken as the spores ripen. The flesh is white and almost odourless when fresh, but on drying it develops a distinct 'mushroomy' or truffle-like smell. It is poisonous.

5 **Inocybe fastigiata** grows especially in beech woods in summer and autumn. The pale yellowish-brown cap is cone-shaped when young, but with age becomes flattened, with a distinct boss (umbo) in the centre. The surface is radially cracked. The long, thin stalk is paler coloured than the cap. The gills are yellow, and the spores brown. The white flesh smells faintly of meal and is poisonous.

6 **Cortinarius elatior** is found in beech woods in late summer and early autumn. The whole plant is rather slimy to the touch. The cap is pale brown and marked with radial grooves. The white stalk is thick and has a short root-like part under the ground; it tapers at the top and at the bottom and is incompletely ringed by greyish-white scales. The pale-brown gills are covered with a cobweb-like veil (cortina) when young, but become darker and sometimes tinged reddish violet with age. The spores are brown. A related species, *C. semisanguineus*, is shown on p. 115.

7 **Tricholoma argyraceum** grows in beech woods on chalky soils, and also amongst conifers, in summer and autumn. The cap is grey, sometimes tinged with brown, and is covered with small, hairy scales. The gills are greyish white when young, but become pale yellow with age; and the greyish-white stalk is usually hollow. The flesh, which has a mealy taste, is edible.

1 AMANITA PHALLOIDES 2 AMANITA CITRINA 3 AMANITA CITRINA *var.* ALBA
4 INOCYBE MACULATA 5 INOCYBE FASTIGIATA 6 CORTINARIUS ELATIOR
7 TRICHOLOMA ARGYRACEUM

Beech woods are mainly confined to the southern half of England, where, although the tree is at a disadvantage compared with oak on heavy clay, it is able to compete successfully on loams and deep sands, and flourishes also on shallow soils, such as those overlying the chalk. Fungi are frequently abundant, but other plants are scarce because comparatively little light penetrates the close canopy of foliage. The bark is thin and smooth and is not very favourable to the growth of lichens, mosses, and liverworts, although some may be found in rain tracks down the trunks and where the larger roots form buttresses at the bases of the trunks.

1 **Lactarius pallidus** grows under beech trees in summer and autumn. The whole plant is a pale yellowish brown, sometimes with a slight tinge of pink. Though slightly domed when young, the cap soon becomes flat and eventually hollow in the centre. The gills are narrow and crowded, and the smooth stalk is rather short and thick. The flesh exudes a white 'milk' when broken, which tastes bitter or very slightly peppery.

2 **Lactarius blennius** is found in woodlands, especially beech woods, in late summer and autumn. The cap is brown with a distinctly greenish tinge, and is patterned with darker spots which form zones on the surface. The stalk is similarly coloured but paler. The white gills turn grey when bruised, and the 'milk', which tastes very peppery, also turns grey on exposure to air.

3 **Mycena capillaris** grows on rotting beech leaves in autumn and early winter. The whole plant is white. The gills are widely spaced and not very numerous, and the stalk is long, thin, and rather limp, with a slight swelling at the base which is covered with short hairs. The white flesh has neither smell nor taste.

4 **Mycena pura** is found in beech and other woodlands from late spring until early winter. The colour of the plant is variable, different shades of lilac or pink being found. The gills are broad, and the stalk is covered with short, woolly hairs at the base. The white flesh may be tinged with pink, and has a mild taste and a strong smell of radishes.

5 **Mycena pelianthina**, which is very much like *M. pura*, usually grows in beech woods. The cap is greyish lilac, and so are the gills, but they are edged with a dark purple line. The flesh is white or tinged violet, and has a mild taste and a smell of radishes.

6 **Mycena vitilis** is found amongst rotting leaves, especially of beech, in summer, autumn, and winter. The grey or brownish-grey cap usually has a raised boss (umbo) in the centre. The gills are very pale grey, and the spores are the colour of cartridge paper. The stiff but elastic stalk is long and thin with a smooth shiny surface, although there may be a few long stiff hairs at the base. It can often be traced to a twig lightly buried in the ground. Other species of *Mycena* are shown on pp. 105, 133, and 145.

7 **Russula olivacea** is variable in colour, the cap being a mixture of brown, dull red, and sometimes drab olive tints. It is usually marked with alternating lighter and darker zones, and is paler at the margin. The white stalk is flushed with pink, at least near the top; if it is soaked in water in which a few crystals of phenol have been dissolved, it turns purplish red. The gills are white when young, but become deep yellow as the egg-yolk-coloured spores develop.

8 **Russula virescens** has a green cap, with a coarsely granular surface which soon becomes cracked, showing the white flesh underneath. The gills and spores are white or very pale cream. The white stalk is short and tapers towards the base.

9 **Russula lepida** has a deep rose-red coloured cap, the surface of which is dusted with a fine white powder. The skin cannot be peeled off. The stalk is similarly coloured, but the gills and spores are white or pale cream. The white flesh tastes slightly bitter and smells faintly of cedar wood.
R. rosea is very similar, but the skin can be peeled from the cap, at least to about half way, and the flesh has no taste or smell.

10 **Russula fellea** has a dull yellowish-brown cap which is slightly slimy to the touch. The gills and stalk are similarly coloured but paler, and the spores are pale cream. The white flesh tastes peppery and smells like Pelargonium. There is a general note on the genus *Russula* on p. 136.

1 LACTARIUS PALLIDUS   2 LACTARIUS BLENNIUS
3 MYCENA CAPILLARIS   4 MYCENA PURA   5 MYCENA PELIANTHINA   6 MYCENA VITILIS
7 RUSSULA OLIVACEA   8 RUSSULA VIRESCENS   9 RUSSULA LEPIDA   10 RUSSULA FELLEA

1 **Craterellus cornucopioides** ('Horn of Plenty') grows in groups in late summer and autumn, and is most often found associated with beech trees. The fungus is funnel-shaped, has a wavy margin, and is brownish black on its inner surface. The spore-producing layer, on the outside, is smooth or slightly wrinkled and greyish black in colour. The spores are white. Its funereal appearance has given it the name '*Trompette des morts*' in France, but in spite of its name and appearance, it is edible when cooked and has an agreeable flavour. It dries easily on account of the thinness of the flesh and its rather tough papery texture, and then can be ground up and used as a flavouring.

2 **Cantharellus cibarius** ('Chanterelle') is found in all kinds of woodland in summer and autumn. It is funnel shaped and egg-yolk yellow in colour; the paler-yellow flesh has a firm texture and a faint but distinct smell of apricots. The spore-producing layer is thrown into narrow folds resembling gills, which branch irregularly and often reunite to form a network. The spores are pale pinkish buff. This fungus is valued in the kitchen because it keeps and cooks well. It can be confused with *Hygrophoropsis aurantiaca* (p. 103), but the latter is orange rather than yellow and does not smell of apricots.

3 **Cantharellus infundibuliformis** grows in groups in all kinds of woodlands throughout the summer, autumn, and early winter. It is funnel shaped, with a wavy margin, a dark brown inner surface, and a long yellow stem. The outer, spore-producing surface is yellow at first but becomes grey with age, and grows in blunt, irregularly-branched folds. The spores are white. The plant is edible when cooked. An entirely yellow variety, *lutescens*, is sometimes found.

4 **Oudemansiella radicata** is found especially associated with beech trees in summer and autumn. The yellowish-brown cap is radially grooved or wrinkled and is slimy to the touch. The gills are white, rather thick, and widely spaced; the spores are white. The long, slender stem is white, tinged with brown, and is hard and tough, with a grooved surface, and narrows towards the top. Below ground it continues as a long, tapering, root-like structure, which eventually contacts wood.

O. longipes, a rare plant of beech woods, is similar to *O. radicata*, but the cap and stalk are dark brown and covered with a dense coat of very short hairs which give them a velvety texture.

5 **Collybia confluens** grows in clusters which are often arranged in rings, usually in beech woods, in summer and autumn. The cap is very pale brown, often tinged with pink; the gills are white, tinged pinkish brown; and the spores are white. The surface of the stalk, which is rather darker than the cap, is covered with a dense coat of short hairs and is flexible and leathery in texture.

6 **Collybia fusipes** ('Spindly Foot') is found from late spring until early winter, growing in groups at the base of beech and oak trees. The clusters arise from a dark coloured underground mass of tissue which is attached to the roots of the tree and persists for many years. The dark reddish-brown cap may be almost liver coloured when wet, but is paler when dry. The stalk is a similar colour, becoming darker brown at the bottom; it tapers both at the top and towards the base, and the surface is deeply furrowed or grooved. The gills are pale reddish brown and often blotched with darker coloured spots; they are rather thick and distantly spaced. The spores are white. The firm, tough flesh has a faint fragrant and a spicy smell, resembling a mixture of lavender and cinnamon. Other species of *Collybia* are described on pp. 103, 127, and 133.

1 CRATERELLUS CORNUCOPIOIDES
2 CANTHARELLUS CIBARIUS   3 CANTHARELLUS INFUNDIBULIFORMIS
4 OUDEMANSIELLA RADICATA   5 COLLYBIA CONFLUENS   6 COLLYBIA FUSIPES

1 **Oudemansiella mucida** ('Beech Tuft') is found in late summer and autumn, growing in clusters on trunks and branches of beech trees, often at a considerable height from the ground. The whole plant is shining white and covered with a sticky fluid (mucilage) which makes it very slimy to the touch. The cap is pure white, or ivory, or sometimes faintly tinged with grey; and the stalk has a large ring which hangs downwards. The gills are broad and rather widely spaced, and the spores are white.

2 **Pleurotus ostreatus** ('Oyster Mushroom') grows on beech and various other trees in late autumn and winter. The cap is deep bluish grey when young, but becomes paler, and finally fawn or brown. The thick white stalk attached to the side of the cap is either very short or may be almost absent. The gills, which are broad, rather widely spaced, and run down the stalk, are white when young, becoming tinged yellow with age. The spores are pale lilac. The flesh has no marked smell or taste, but since the fungus, which often grows in clusters, can be collected in quantity and dries very well for storage, it is usually listed as edible, though of moderate quality. It causes a destructive rot of the wood in the trees on which it grows.

3 **Ganoderma applanatum** is found on a wide variety of trees, being particularly common on beech. The upper surface is greyish or reddish brown, with a hard crust which, when cut, has the appearance of freshly-broken resin. The tubes on the lower surface are white when young, but become pale brown with age; they stain a darker brown when bruised. The brown spores are produced very copiously and are frequently to be seen deposited in quantity on the upper surface and on nearby branches and any projections providing a lodgement. This fungus is perennial and causes a serious heart rot in the trees which it attacks.
G. lucidum is an annual species growing on oak and other trees. The upper surface has a thick, resinous crust, and is shiny and dark brown or almost black, sometimes with a purple tinge. The tubes are whitish brown and have very small openings.

4 **Gloeoporus fumosus** is found in summer, autumn, and early winter on beech and many other trees. It grows in groups, forming crowded tiers of brackets. The upper surface is pale brownish or greenish grey and is closely covered with very short hairs; the lower surface is pale grey when young, becoming greyish brown with age. The greyish white flesh is tough and flexible and has a distinct sweetish sickly smell; it is separated from the layer of tubes on the underside by a thin grey or almost black jelly-like zone which appears as a dark line when the fungus is cut or broken. This species is as widely distributed but not as common as *Gloeoporus adustus* (p. 157).

5 **Deedalea quercina** grows on oak and sweet chestnut as well as beech, and occasionally on other trees. The upper surface is pale greyish brown when young, becoming darker with age. The 'tubes' on the lower surface are pale brown, and the openings are considerably elongated and twisted, somewhat resembling a maze (daedaloid). This fungus is perennial, but it does not appear to attack living wood, although it has been seen on dead branches at a considerable height from the ground.

Lenzites betulina, though found chiefly on birch, is quite frequent on other trees. The upper surface is pale greyish brown, with alternating lighter and darker zones, and is covered with short, woolly hairs. The spores are produced on the lower surface on yellowish-grey, radially-arranged, branching plates, which often resemble gills; in some specimens the cross connections are numerous enough to produce a daedaloid appearance, as in No. 5.

6 **Schizophyllum commune** grows on dead branches of other trees as well as birch. Usually it consists of a fan-shaped cap attached by a short stalk; sometimes, however, there is no stalk, and occasionally when it is growing on the underside of a branch it is attached by its upper surface. The gills on the under surface are pale grey with a distinct violet tinge; they are split longitudinally, which shows most conspicuously in dry weather, when the two halves of each gill twist slightly away from each other. The upper surface is grey when moist but almost white when dry, and is covered with very short soft hairs. This fungus is quite common in southern England, especially in the south east, but is rare elsewhere.

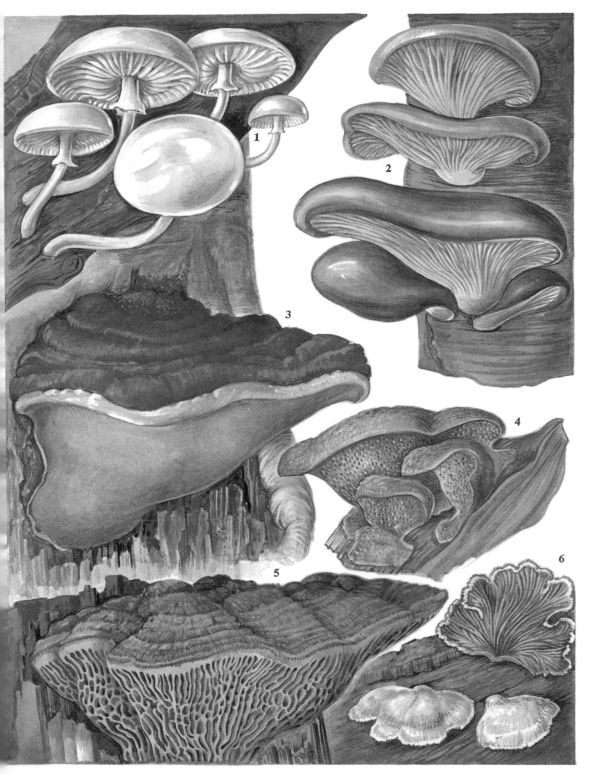

1 Oudemansiella Mucida   2 Pleurotus Ostreatus
3 Ganoderma Applanatum   4 Gloeoporus Fumosus
5 Deedalea Quercina   6 Schizophyllum Commune

Oak woods are the most widespread natural plant communities throughout England, and in the valleys of Wales and Scotland. If the trees grow close together, the canopy of foliage is quite thick, although not so dense as that of beech, and this restricts the amount of light beneath the trees. But most woods are regularly thinned, and then they have a well-developed shrub layer and a rich community of other plants. Ferns and mosses flourish in the ground flora; fungi are common. Oak-tree bark is rugged and not very acid in reaction, and lichens, mosses, and liverworts grow very well on it.

1 **Lactarius quietus** is found in late summer and autumn on the ground near oak trees. The cap is dull reddish brown, with alternating lighter and darker zones; the stalk is similarly coloured and marked with uneven furrows. The gills, pale at first, become darker with age. The whole plant has a faint, sweetish, oily smell, said to resemble that of cockroaches. The pinkish brown flesh, when broken, exudes a small amount of cream-coloured or very pale yellow 'milk' which tastes mild at first, but produces a slight burning sensation after a few moments. It should not be eaten.

2 **Collybia dryophila** grows in woodlands, especially oak, from late spring until early autumn. It has a yellowish-brown cap, and a long, slender, tough and flexible stalk, which is swollen and often tinged reddish brown at the base. The white or pale yellow gills are narrow and crowded together. The spores are white. Other somewhat similar species of *Collybia* are shown on p. 133.

3 **Tricholoma sulphureum** is found associated with oak trees in autumn and early winter. The whole plant is bright yellow. The cap is often rather irregular in shape and splits easily at the edges; the surface is silky to the touch. The gills are thick and widely spaced, and the spores are white. The yellow flesh has a strong distinctive smell like that of coal gas.
A similar species of pine woods, *T. flauovirens*, is greenish yellow rather than pure yellow and lacks the characteristic coal gas smell of *T. sulphureum*.

4 **Mycena inclinata** grows in tufts on the stumps of oak and sweet chestnut trees in late summer and autumn. The cap is brownish grey, with a much paler margin which is usually toothed above the gills. The long slender stalk is white at first, but later becomes reddish towards the base; it is silky to the touch, often twisted and elastic, but not tough. The flesh has a strongly rancid smell.

5 **Mycena polyadelpha** is found on rotting oak leaves in autumn. The cap is very delicate, pure white, and grooved; the long, thin stalk has a yellowish tinge. Other species of *Mycena* are described on pp. 33, 105, 133, and 145.

6 **Russula sororia** has a greyish-brown cap, which is darker in the centre and rather thin at the margins. The gills are narrow and crowded, and the white, spongy flesh has an acrid taste and a rancid smell.

7 **Russula vesca** has a pinkish-brown cap which usually has a slight hollow in the centre. In old specimens the flesh is often exposed as a narrow zone at the margin, and the narrow, crowded gills may become blotched with brown spots. The flesh, which turns pale yellowish brown when broken, is odourless but has a mild and nutty taste. There is a general note on *Russula* on p. 136.

1 Lactarius Quietus   2 Collybia Dryophila   3 Tricholoma Sulphureum
4 Mycena Inclinata   5 Mycena Polyadelpha
6 Russula Sororia   7 Russula Vesca

1 **Polyporus squamosus** ('Dryad's Saddle') is found from spring onwards, sometimes on stumps but often either high up on the trunk or near the ground, on living trees. It grows on oak, maple, and other kinds of trees, but especially on elm. It is a large fungus (Fig. 1A shows a typical specimen one-tenth natural size), and its colour, horizontal growth, and general shape, with the strongly incurved margins, are reminiscent of a saddle. The ochre-yellow upper surface is ornamented with flat, feathery, dark-brown scales arranged in more-or-less concentric rings. On the under surface the pores are white, with a rather dirty yellowish tinge in older specimens. The firm, white flesh becomes tough with age, and has a faint, pleasant smell, usually described as 'sweetish'; but the taste is strong and unpleasant. When *P. squamosus* becomes established on a living tree, it causes a destructive white rot of the wood.

Polyporus varius is somewhat similar to *P. squamosus* but smaller. The brownish-yellow upper surface lacks scales, although it may be streaked with radiating dark-brown lines. The flesh is yellow and very tough.

Polyporus brumalis; P. nummularius. The stalks of these species are placed centrally. They sometimes grow from buried sticks or pieces of wood, and so appear to be growing on soil, when they can be mistaken for a species of *Boletus*. However, the 'cap' is quite different, being very thin and extremely tough, and the tubes are not separable from it. *P. brumalis* has a greyish-brown 'cap' and a pale-grey stalk; *P. nummularius* has a pale brownish-yellow 'cap', and the stalk is dark grey, almost black.

2 **Fistulina hepatica** ('Beefsteak', '*Langue de boeuf*') grows on living trees, usually oak or sweet chestnut, in late summer and autumn. It reaches its full size in about 2 weeks, after which it decays rapidly. The dark blood-red upper surface is sticky and slimy; it has a velvety appearance and is slightly rough to the touch due to a dense covering of short tufted hairs. The short very narrow tubes on the under surface are closely packed, although not joined to one another, and their pale yellow colour later becomes flushed with red. Although Schaeffer, writing in 1780, said that it was used for food by poor people in Bavaria, its common names refer to its appearance rather than to its edibility, for it certainly resembles meat in many ways. It is often tongue-like in shape as well as colour

and surface texture; the upper surface moves when pressed, like skin on a body; the flesh is thick, with a fibrous texture like the poorer cuts of steak, and it yields red juice freely when cut. But, in spite of its pleasant fruity smell, its taste is sour and acrid. Even when cooked, with the addition of onion or garlic and mixed herbs, it is not very palatable. *F. hepatica* causes a brown rot of the wood of its host tree, causing the timber to become a rich warm brown without a great loss of strength, at least in the earlier stages. Such timber used to be much prized by cabinet makers.

3 **Grifola sulphurea** is found usually on oak trees but sometimes on other trees, including conifers, from late spring onwards. It is fan-shaped, several 'fans' sometimes growing in tiers from the same stalk. The orange-yellow or flame-coloured upper surface fades to a yellowish tan, like pale leather, and on the under surface the very small pores form a thin, bright sulphur-yellow layer. The flesh is thick, pale yellow, and somewhat spongy when the fungus is young, but developing a cheese-like texture later; it has a faint but distinct sour smell, and the taste is sour and acrid. This fungus causes a destructive red rot of the heart wood of its host tree.

4 **Grifola gigantea** grows from the summer through the autumn and winter at the base of oak trees, and also of beech and sometimes other trees. It is fan-shaped, the 'fans' usually growing in tiers and groups and often forming a mass of considerable size (Fig. 4A shows an average specimen one-tenth natural size). The brown or leather-coloured upper surface is densely covered with small, brown, fluffy scales, and the very small pores on the under surface form a thin white layer. Both the white flesh and the pores turn brownish red and then black when bruised or cut. Although young specimens are somewhat brittle, the mature fungus is tough in texture; it has a faint sour smell and taste.

Grifola frondosus. The densely-tufted narrow 'fans', grey-brown in colour, can be said to resemble a bundle of feathers, and in Canada this species is called 'Hen of the Woods'. The rather soft white flesh smells of acetamide (like the smell of house mice). It is not common in Britain and is most often found growing at the bases of oak trees.

A rare species, *G. umbellata*, differs in that the stalks of the 'fans' are centrally attached.

1 Polyporus Squamosus   2 Fistulina Hepatica
3 Grifola Sulphurea   4 Grifola Gigantea

Almost all woodlands in Britain are the result of planting over a long period of time with varying degrees of foresight, and nowadays few trees are allowed to grow to their full size. This has resulted in woodland flowerless plants becoming much scarcer than they were 100 years ago; it is also not possible to state with any accuracy the associations of species that are likely to occur. In general, however, chestnut coppice is perhaps the least rewarding type of woodland, and of all trees elder supports the most varied population of mosses, liverworts, and lichens.

1 **Lactarius vellereus** ('Velvet Cap') is found in mixed woodland, often growing in groups, in the autumn. The cap is white, with a velvety surface; the stalk is also white and rather short. The gills, which are thick and widely spaced, are white at first but soon become a pale ochre colour. When broken, the flesh exudes a white latex (milk), which tastes very peppery and also turns orange-brown when mixed with a drop of potassium hydroxide solution.
All *Lactarius* species have a brittle flesh and exude a milky latex when broken. They have white or cream-coloured spores.

Lactarius piperatus is another white species which is fairly common. But the cap is smooth, the gills are very crowded, and the milk, although very peppery to the taste, does not change colour with potassium hydroxide solution.

2 **Lactarius mitissimus** is found in the autumn in all kinds of woods, often associated with conifers. The cap and stalk are orange, and the gills are similar but paler. The milk is slightly bitter.

3 **Lactarius pyrogalus** is chiefly associated with hazel and appears in the autumn. The cap is greyish or yellowish brown, and the stalk is similar, but paler. The gills are yellow at first, becoming dark ochre later. The milk is very peppery, and turns orange when mixed with a drop of potassium hydroxide solution. *L. circellatus* is an uncommon plant associated with hornbeam. It is rather like *L. pyrogalus*, but the peppery tasting milk is unchanged by potassium hydroxide solution.

4 **Lactarius subdulcis** is a common species of mixed woodland, often associated with beech, and to be found from late summer to autumn. The cap is buff or tan with a distinctly reddish tint (often paler than the specimen illustrated here); the similarly-coloured stalk is paler at the top, darkening towards the base. The gills are buff with a flesh tint. The milk has a slightly bitter taste. Other species of *Lactarius* are illustrated on pp. 105, 113, 121, and 127.

5 **Amanita virosa** ('Destroying Angel') is a rare plant of mixed woodland, often on rather poor soil. It is very poisonous; its earliest known victim was perhaps the Emperor Claudius. It occurs in North America as well as in Europe and has caused many deaths in both continents. The symptoms produced are similar to those caused by the deadly *A. phalloides* (p. 119). It develops at first completely enclosed within a white membrane, and when later this bursts, it remains as a cup at the base, called a volva. The cap, stalk, and gills are pure white. It is deceptively innocent in having no distinctive colouring, taste, or smell, and not changing colour when bruised or peeled. It can be distinguished from edible mushrooms by the white gills (the edible kinds have grey, pink, or dark brown gills) and by the volva (not present in mushrooms). Also, the flesh of the cap turns bright yellow when treated with potassium hydroxide solution. If in doubt, and the 'mushroom' is growing in a wood rather than in pasture, avoid it.
The illustration shown here is after M. C. Cooke, *Illustrations of British Fungi*, 1880-90.

6 **Amanita rubescens** ('Blusher') is found commonly in all kinds of woodland, including coniferous, in summer and autumn. The cap is a dull reddish brown or tan (said to resemble the skin of roast goose), with rather dirty white patches. The stalk and gills are white, and the bulbous base of the stalk usually has several rows of warty patches which are the remains of the volva. Though not poisonous, it is indigestible and has an unpleasantly sweet taste if eaten raw. It can be distinguished from its poisonous relatives by the way in which the flesh reddens when bruised: in mature specimens the gills have red spots on them, and the stalk is flushed red at the base.

Amanita pantherina ('False Blusher'). A somewhat similar species in appearance, but poisonous, and found especially in beech woods in late summer and autumn. The cap is greyish brown with white patches. The stalk and gills are white and do not redden when bruised. The flesh of the cap turns orange-yellow when treated with a drop of potassium hydroxide solution.

7 **Clitocybe odora** ('Anise Cap') is found in mixed woodland in late summer and autumn. The whole plant is blue-green in colour, and has a fragrant odour of anise; this persists when the fungus is dried, when it may be used as a flavouring.
C. fragrans is a pale yellowish-brown fungus; and *C. suaveolens* is darker brown and is found in conifer woods. Both have a similar fragrant anise odour. Other species of *Clitocybe* are shown on pp. 35 and 107.

1 LACTARIUS VELLEREUS

2 LACTARIUS MITISSIMUS 3 LACTARIUS PYROGALUS 4 LACTARIUS SUBDULCIS
5 AMANITA VIROSA 6 AMANITA RUBESCENS 7 CLITOCYBE ODORA

1 **Laccaria laccata** ('Deceiver') is found from summer to winter in woodlands, and also sometimes in the open in heathy and boggy places. It varies considerably in size and in the shape of the cap, which may be convex, or flattened, or raised in the centre, or slightly depressed. The colour, though generally reddish brown, also varies, sometimes being brick red when wet, and drying to a ruddy flesh colour or a paler yellowish red. The cap may be either somewhat scaly in the centre, or smooth. The flesh-coloured gills, are thick and distantly spaced, and become conspicuously powdered with the white spores. The fibrous and rather elastic stalk is covered with white, short, woolly hairs at the base.

2 **Laccaria amethystea** ('Amethyst Deceiver') closely resembles *L. laccata* in form and structure and has been considered to be merely a variety of it. The two species often grow together, although *L. amethystea* is rather less common, is confined to woodlands, and has a slightly shorter season. The whole plant is an amethyst or deep violet colour, which becomes paler on drying.

3 **Collybia peronata** ('Woolly Foot') grows amongst leaf litter in woodlands in late summer and autumn. The reddish-brown cap fades to yellowish brown; and the yellow gills also become brown with age. The spores are white. The stalk is tough (cartilagenous), and yellowish brown in colour, and its curved base is clothed with a pale yellowish-white 'wool' which adheres to dead leaves and small twigs on the surface of the ground. The flesh is leathery and pliant and tastes peppery when chewed. Other species of *Collybia* are shown on pp. 103 and 123.

4 **Collybia butyracea** ('Greasy Club Foot') is found in all kinds of woodland in autumn and early winter. The brown cap, which is greasy to the touch has a distinct boss (umbo) in the centre. The gills are greyish brown, and the spores are white. The tough (cartilagenous) stem tapers upwards from an enlarged spongy base, resembling that of *Clitocybe clavipes* (p. 107), which grows in similar places. However, although the cap of *Collybia butyracea* may become grey on drying, the umbo always remains brown, and the gills do not run down the stem as they do in *Clitocybe clavipes*.

5 **Mycena hiemalis** ('Fairy Button') grows on pieces of wood or on the bases of trees, often amongst mosses, and though it may be found at all times of the year, it is frequently seen in winter. The cap, which is pale brownish grey, is conical when young, but later the margin turns upwards and becomes prominently grooved. The dingy white gills are somewhat distantly spaced, and the spores are white.
A very similar species is *M. olida*, but it is considerably less common. It has a white cap and crowded gills.

6 **Mycena tenerrima** ('White Bonnet') is found on wood or on the trunks of trees at all times of the year. The tiny white cap is finely powdery with a frosted appearance, and is steeply conical in shape. The gills are white and crowded, and the spores are white. *M. tenerrima* can be confused with *M. corticola* (p. 145) which often grows in similar situations.

7 **Mycena acicula** ('Orange Bonnet') grows on small pieces of wood from late spring until autumn. Its steeply conical vermilion-tinged cap is orange, tinged with vermilion; the gills are yellow and distantly spaced, and the spores are white.

8 **Mycena galericulata** ('Grey Bonnet') is found on wood, particularly on old stumps, at all times of the year. The cap is white tinged with pale grey or fawn; it is conical at first, becoming flattened with a distinct central boss (umbo) later. The gills, which are white when young, become pink with age, and are connected by cross veins at the base. The spores are white.

9 **Mycena galopus** ('Milk Drop') grows on small pieces of wood and twigs amongst dead leaves from late summer until early winter. The cap is grey, conical, and strongly grooved, the gills greyish white, and the spores cartridge-paper coloured. The grey stalk is hairy at the base and yields a milky latex when broken. A pure white variety *candida* is sometimes found.

Mycena leucogala, which is similar to *M. galopus* in shape and form and in the white latex, has a very dark brown or almost black cap. It is found on burnt ground. Two other species *M. sanguinolenta* and *M. haematopus*, yield latex, but it is a dark blood-red instead of white. The former has a pale brownish-red cap, often tinged purple, and flesh-coloured gills edged with red; the latter has a greyish-brown purple-tinged cap and white gills tinged pink or lilac. It grows in clusters.

10 **Mycena polygramma** ('Roof Nail') is found on wood, often on stumps or fallen branches, in summer and autumn. The cap is steel grey in colour and conical in shape, with a pronounced central boss (umbo). The long steel-grey stalk is hairy at the base and marked with raised spiral lines. The gills are greyish white, and the spores the colour of cartridge-paper. Other woodlands species of *Mycena* are on pp. 105, 121, and 145, and grassland species on p. 33.

1 Laccaria Laccata   2 Laccaria Amethystea   3 Collybia Peronata   4 Collybia Butyracea
5 Mycena Hiemalis   6 Mycena Tenerrima   7 Mycena Acicula
8 Mycena Galericulata   9 Mycena Galopus   10 Mycena Polygramma

1 **Pluteus cervinus** ('Deer Toadstool') is found at all times of the year, frequently growing on old stumps and fallen trees, but also on wood chips and sawdust. The cap is sooty brown and slimy to the touch when wet, and it comes off the stalk very easily. The gills are white when young, but become salmon coloured as the pink spores develop. The white stalk has brownish-grey longitudinal streaks. The flesh has a faint but distinct smell, resembling that of radishes, and is edible though it has a somewhat sour taste when raw.

About thirty other species of *Pluteus* have been recorded in Britain, and can be recognized by the absence of a ring on the stem or a 'cup' or volva at its base, the pink spores, and easily separable cap. None of them is at all common.

Volvariella speciosa has a greyish-brown cap, and gills which are white when young but later become pink, like the spores. There is a 'cup' or volva at the base of the stem. *V. speciosa* is occasionally found on compost heaps or amongst rotting grass or straw, and is good to eat. A related species, *V. volvacea*, is one of the very few fungi which are cultivated for food. The variety *masseei* has been cultivated on composted straw in India for a very long time and is known as the 'Padi Straw Mushroom'; the variety *heimii* is similarly cultivated in Madagascar.

2 **Lepista nuda** ('Blewits') grows on rich soil in woodlands and gardens in late autumn and early winter. The cap is bluish lilac when young, but becomes brownish violet with age; the stem, which is similarly coloured, has a mealy texture. The pale lilac gills are crowded and become tinged with brown in older specimens. The pale violet flesh has a pleasant smell and is good to eat. This fungus has been called *Tricholoma nudum*, but the spores are pink, not white. (*See* also p. 31.)

Lepista irinum is like *L. nuda*, but is pinkish buff in colour, with no trace of violet. The flesh is white, and has a strong fragrant perfume resembling that of orris root. It is not common anywhere, but is perhaps most often found in East Anglia.

3 **Tricholoma saponaceum** ('Soap-Scented Toadstool') is found often growing amongst mosses in late summer and autumn. The cap is greyish brown, with a tinge of olive green, and has a few small scales in the centre. The gills are often more buff coloured than in the specimen shown and have bluish grey margins with sometimes a few scattered red spots; the spores are white. The stalk, which is also buff with darker streaks, is thickest near the bottom and tapers upwards, while the base terminates in a short point, almost like a root. The white flesh turns faintly pink when broken; it smells and tastes distinctly of kitchen soap.

Tricholoma gambosum ('St. George's Mushroom') grows in undergrowth on the edges of woods, amongst light scrub on hillsides, and in open grassland, especially on chalky soils, in the spring. It usually first appears about St. George's day, and is at its most plentiful in the following month, although it is never very common. The cap is convex when young, but becomes flattened, with wavy margins, later, and the stem is rather thick and fleshy. Its colour is predominantly creamy white, becoming a very pale buff with age. The crowded gills and spores are white. The flesh is white, with a strong smell of fresh meal and a pleasant taste when eaten. Other species of *Tricholoma* are shown on pp. 115 and 119.

4 **Russula foetens** is yellow, tinged with brown or grey. The gills often become spotted brown and exude drops of moisture. The flesh has a bitter peppery taste and a strong oily smell.

5 **Russula nigricans.** Though white when very young, this species gradually becomes sooty brown and eventually black. When fresh, the flesh turns pink if it is cut or broken; it also turns dark olive green when rubbed with a crystal of iron sulphate or iron alum.

6 **Russula cyanoxantha** is characteristically coloured a variety of lilac, purple, and green. The gills are elastic and oily to the touch. The flesh does not change colour, apart from sometimes becoming faintly olive green, when rubbed with a crystal of iron sulphate or iron alum. There is a note on *Russula* on p. 136.

1 Pluteus Cervinus   2 Lepista Nuda   3 Tricholoma Saponaceum
4 Russula Foetens   5 Russula Nigricans   6 Russula Cyanoxantha

Nearly a hundred species of the genus *Russula* have been recorded in Britain, though many are rare. Twenty-four of the commoner species are illustrated in this book, on pages 107, 115, 121, 127, 135 and 137, and several others are mentioned in the text. They are conspicuous plants of the woodland floor in late summer and autumn, and are usually brightly coloured. They all have domed caps which later become flattened, and finally depressed in the centre. There is no ring on the stalk nor 'cup' or volva at the base. The flesh is brittle and easily broken, like that of *Lactarius*, but the latter has a milky latex which is never found in *Russula*. In most species the stalks and gills are white, and there are no intermediate 'short' gills. The spores are white; and the flesh turns pink when rubbed with a large crystal of iron sulphate or iron alum. There are exceptions to all these statements, which are very useful in identification, and so they have been pointed out in each case where they occur.

1 **Russula ochroleuca** is bright yellow or ochre coloured. The flesh has a very faint, pleasant smell, and tastes mild or slightly acrid.

Russula solaris is also yellow, but is less common, and is confined to beech woods. The gills become straw coloured with age, the spores are deep cream, and the flesh tastes distinctly acrid. *R. smaragdina* is a pale lemon yellow (though the word 'smaragdine' means emerald green), and is a rare plant of oak woods.

2 **Russula azurea** is a pale grey, mixed with blue and lilac tints, and the surface of its cap is covered with small white granules. It is a rare species, associated with conifers.

3 **Russula emetica** is pure scarlet with no purple tint. The skin can be peeled from the cap, and the pink flesh is very acrid to the taste and may cause vomiting if eaten in quantity. It is associated with conifers. *R. mairei* is a very similar species found growing under beech trees. It peels less easily, and the flesh is white with a faint honey smell.

4 **Russula fragilis** is crimson tinged with violet or sometimes quite mauve, as the specimen illustrated. It has a dark centre. The flesh has a pleasant smell and tastes distinctly peppery.

5 **Russula delica** is a rather dirty-white colour, and the gills are tinged bluish-green where they run down the stalk. There are intermediate short gills. The stalk has a mild taste, but the gills are acrid. The plant has a faint but distinctive smell said to be 'compounded of ivy and bitter orange'.

6 **Russula xerampelina** varies from a purplish red or wine colour to a purplish brown, and the spores are ochre coloured. The flesh smells faintly but distinctly of lobster and turns deep olive green when rubbed with a crystal of iron sulphate or iron alum.

Russula queletii is also purplish red or wine coloured, and the stalk is flushed with the same colour as the cap. The flesh is acrid tasting and turns pink when rubbed with a crystal of iron sulphate or iron alum. It is associated with conifers, as is also the similar *R. sardonia*, which is described on p. 106.

7 **Russula atropurpurea** is a dark purple-red colour with a black centre, and a stalk tinged rusty-brown at the base. It has a mild taste and is not acrid or peppery like *R. Fragilis*.

8 **Russula violeipes** is lemon-yellow, sometimes tinged pink; the stalk, and sometimes the rest of the plant, is flushed violet. The flesh has a characteristic fruity smell.

1 Russula Ochroleuca   2 Russula Azurea   3 Russula Emetica   4 Russula Fragilis
5 Russula Delica   6 Russula Xerampelina
7 Russula Atropurpurea   8 Russula Violeipes

1 **Coprinus lagopus** is found in mixed woods in summer and autumn, usually growing singly amongst leaf litter on the ground, or sometimes on small twigs. When it first appears the cap is thimble shaped, but it soon becomes flat, and finally the margins turn upwards. It is white and densely covered with hairs which are also white at first, but which darken later and eventually fall off. The gills are dark grey, and as the spores are shed, they wither until they finally appear as mere ridges on the underside of the cap. The white stalk is very hairy, especially towards the base, and elongates considerably as the fungus grows.

2 **Coprinus micaceus** grows in clusters on or by the stumps of a variety of trees at any time from late spring onwards. The cap is brown, often date brown, but sometimes a dark ochre colour; it is deeply grooved and when young is sprinkled with glistening particles which often disappear in older specimens. The dark-brown gills break down to some extent to form a dark liquid containing the spores, but this process is not so marked as it is in some other species of *Coprinus*. The stalk is white and is densely covered with very short stiff hairs.
C. truncorum is similar, but the stalk is smooth and quite free from hairs. *C. silvaticus* also resembles *C. micaceus*, but the stalk has a brownish tinge, and the covering of hairs is softer. The cap and gills break down to form a dark liquid containing the spores.

3 **Coprinus disseminatus** ('Crumble Cap'). Large clusters of this tiny fungus are found from late spring onwards in mixed woodland, often on rotten stumps. The pale yellowish-brown cap is thimble-shaped at first but becomes greyer and more cap-shaped as it matures; it is very thin and deeply grooved. The gills, white at first, become pale grey later, and the stalk is thin, white, and densely covered with very short soft hairs. When the spores have been shed, the whole plant withers and does not break down into a liquid. Other species of *Coprinus* are described on page 34.

4 **Inocybe pyriodora** grows in woods of all kinds in summer and autumn. The pale-ochre cap tinged with reddish brown has a pronounced boss (umbo) in the centre; the surface is covered with small fibrous scales, and the margin is often split. The gills, which are white at first and covered with a delicate veil, become cinnamon-brown later; the upper part of the stalk is covered with a white bloom. The whole plant has a strong smell of over-ripe pears. All species of *Inocybe* appear to be poisonous to a greater or lesser extent, and should be avoided.

Inocybe eutheles and **Inocybe lacera**. These are similar to *I. pyriodora* and both are associated with conifers. The former is a pale fawn colour and smells faintly earthy. The latter, which grows on sandy soils on heaths or in woods, is a snuff-brown colour and smells of fresh meal.

5 **Inocybe geophylla** is a species of damp woodlands and is found in summer and autumn. The white cap, with its central boss (umbo) has a shining silky appearance. The gills are cream coloured and covered with a thin veil at first, but later become a pale brownish grey. The stalk is white and smooth. A lilac-coloured variety, *lilacina* (the specimen shown at the back of the group in the illustration), is common. This fungus is poisonous. Other species of *Inocybe* are shown on page 35.

6 **Psathyrella hydrophila** ('Brittle Cap') grows in tufts on or by stumps of a variety of trees in late summer and autumn. The cap is date brown when moist, but dries out to a pale greyish brown. The gills are pale brown at first, becoming darker. They are covered by a conspicuous veil, fragments of which usually remain on the margins of the cap in older specimens. The stalk sometimes shows a faint ring, which can be seen in some of the specimens in the illustration.
A less common but very similar species is *Psathyrella spadicea*, in which there is no veil over the gills.

7 **Psathyrella gracilis** is found amongst leaf litter and sticks in mixed woodland in late summer and autumn. The cap, which is grooved at the margins, is dark brown when moist, but dries to a pale tan colour. The gills are pale brown with pink edges, and there is no veil. The stalk is long and stiff and has a short rootlike extension below ground.

8 **Psathyrella candolleana** grows, usually in tufts, on or near stumps of various trees from late spring to autumn. The cap is a very pale yellow. The gills are pale grey with a tinge of lilac at first but becoming browner. They are covered by a conspicuous veil, fragments of which remain as toothlike projections on the edge of the cap of older specimens. The slender stalk is easily broken.

1 COPRINUS LAGOPUS   2 COPRINUS MICACEUS   3 COPRINUS DISSEMINATUS
4 INOCYBE PYRIODORA   5 INOCYBE GEOPHYLLA
6 PSATHYRELLA HYDROPHILA   7 PSATHYRELLA GRACILIS   8 PSATHYRELLA CANDOLLEANA

1 **Armillaria mellea** ('Honey Agaric') is found from summer onwards, usually growing in large tufts on stumps and dead trees and at the foot of living trees of all kinds, including conifers. The yellow-brown to deep brown cap is covered with darker hairy scales, which project at first but lie flat later, and eventually disappear from the margin inwards. The gills are cream coloured at first, often with a faint pinkish tinge, and later become yellowish brown with scattered darker spots. The spores are cream coloured. The stalk has a swollen base and a well-developed thick whitish ring with yellow flecks near the top. Below the ring the stalk is covered with an olive-yellow or honey-coloured down. The flesh is white, with a faint, slightly unpleasant smell and a bitter taste, which disappears on cooking. Though the fungus may be eaten, it is only of medium quality. It varies in colour, size, and shape, and is very widely distributed, being common throughout the temperate regions of the northern hemisphere, and also in Australia. It is a serious pest as it causes a most destructive rot in all kinds of trees. It spreads not only by means of spores but also by branching root-like cords, which are about the thickness of a shoe-lace, and dark brown outside and white within. They are found in the soil and under the bark of affected trees.

There is only one other species of *Armillaria*, the rare *A. tabescens*, which has no ring on the stalk.

2 **Galerina mutabilis** grows in large clumps on the stumps of a variety of trees, and is found at any time of the year from spring onwards. The dark-brown cap is slightly sticky when damp; it tends to dry out from the centre and then becomes a much paler ochre colour, with a darker band at the margin. When it is completely dry, the whole cap is pale. The gills are a pale cinnamon colour, becoming darker with age; and the spores are a rusty brown. The stalk has a well-developed ring, below which it is covered with dark brown scales. *G. mutabilis* may be eaten cooked in a variety of ways; it is of good quality, imparting a rich-brown colour, as well as an agreeable flavour, to soups and stews. This fungus is sometimes called *Pholiota* or *Kuehneromyces mutabilis*.

3 **Flammulina velutipes** ('Velvet Foot') first appears in late autumn, and in mild winters may continue until the spring of the following year. It grows on stumps and on the trunks and branches of living trees, and sometimes on old gorse bushes and other kinds of shrub. The bright yellow cap has a darker centre with occasionally an orange tinge, and the pale-yellow gills become darker with age; the spores are brown. The stalk is dark brown and densely velvety, except at the top where it is smooth and yellow. The flesh is pale yellow, and its faintly acrid taste and smell disappear on cooking. Its main advantage as a food is that it is available when other fungi are unobtainable, but it has a somewhat watery flavour.

4 **Hypholoma sublateritium** ('Brick Tuft') is found in late summer and autumn growing in large tufts on stumps and at the base of various kinds of trees. The brick-red cap is paler at the margins than in the centre. The gills are yellow when young, but become greyish violet with age; and the spores are purplish black. Remains of the veil are usually to be seen on the margins of the cap, forming a ring-like zone on the stalk. The flesh is pale yellow and has a faintly bitter taste.

5 **Hypholoma fasciculare** ('Sulphur Tuft') grows in large clumps throughout the year on the stumps of all kinds of trees, including conifers. The cap is bright sulphur yellow tinged with brown in the centre. The yellowish-green gills turn olive or dark brown later, and the spores are purplish black. The remains of the veil can be seen on the stalk as a faint ring-like zone and traces are sometimes left on the margins of the cap. The yellow flesh has an unpleasantly bitter taste.

Hypholoma capnoides is like *H. fasciculare*, but the cap is a brownish-yellow colour, and the flesh does not taste bitter. It is only found associated with conifers.

1 Armillaria Mellea   2 Galerina Mutabilis
3 Flammulina Velutipes   4 Hypholoma Sublateritium   5 Hypholoma Fasciculare

About fifty species of *Boletus* have been recorded in Britain; fourteen of the commoner species are illustrated in this book on pp. 109, 115, and 143, and there are others described in the text. All of them, like *Polyporus* (p. 128), produce their spores in tubes which are packed vertically in a layer beneath the cap; but they differ in that the tubes are easily separated from the flesh of the cap, and the two kinds of fungus are not closely related. The spores are always some shade of brown, in most cases olive brown, like the colour of a bronze medal. The majority are found in woodlands in late summer and autumn. The specimens shown on this plate are of average size, but all these species may grow considerably larger, occasionally even twice as large.

1 **Tylopilus felleus** has an ochre-coloured cap with a slightly velvety surface. The tubes are white at first, becoming pink with age, and the spores are pinkish fawn. The pale olive-brown stalk has a wide-meshed network of darker raised veins on the surface. The flesh is white, but becomes pink when broken, and has a very bitter taste. This fungus used to be included in the *Boletus* genus.

2 **Boletus piperatus** has a smooth, slightly slimy, cinnamon-coloured cap with rather large coppery-brown tubes. The stalk is cinnamon coloured, and at the base the threads of the underground part of the fungus are bright yellow. The flesh, which has a very acrid taste, is mainly yellow, but in the cap it is streaked with red, somewhat resembling rhubarb.

3 **Boletus edulis** (*Cèpe*, *Steinpilz*) is found rather later in the year than the other species. It has a brown, shining cap, and the tubes, white at first, become greenish yellow with age. The stalk is pale brown with a close-meshed network of slightly-raised white veins at the top. The white flesh, slightly tinged with pink, has a rich, nutty flavour, which is not lost when the fungus is dried, and so it is one of the best edible fungi.

4 **Boletus reticulatus,** which appears in summer, has a brown cap with a slightly velvety surface. The tubes are very like those of *B. edulis*, but the spores are a paler colour. The whole stalk has a network of white veins. The flesh is white, tinged pink, and is good to eat.

5 **Boletus erythropus** has a brown cap and greenish-yellow tubes with blood-red or orange openings. The stalk is yellowish brown stippled with red dots. The yellow flesh rapidly turns deep blue (or sometimes red in the stem) when broken; it remains yellow, however, at the base of the tubes.

Boletus luridus is a much less common species, but distinguished by the wide-meshed network of red veins, rather than a stippling of dots, on the stem. The apricot-yellow flesh also rapidly turns deep blue (sometimes with red patches) when broken, but is purplish red at the base of the tubes.

Boletus satanas also turns blue when broken. The cap is greenish grey, the tubes green with red openings, and the stalk yellowish brown with red veins. *B. satanas* causes vomiting when chewed, and so is best regarded as poisonous, although it is not as dangerous as its name implies. It is not common.
B. calopus is another uncommon species similar to *B. satanas*, but the yellow tubes do not develop red openings. Though very bitter in taste, it does not cause vomiting.

6 **Boletus chrysenteron** has a reddish-brown cap, often with an olive tinge, yellow tubes, and a dull red stalk, which is yellow at the top and white at the base. The creamy-yellow flesh is flushed red immediately beneath the skin of the cap, which is frequently cracked so that the red colour shows through. The flesh turns pale blue when broken and later changes to reddish buff. *B. rubellus* is a very similar species to *B. chrysenteron*, but the cap is blood-red or reddish-purple.

7 **Boletus subtomentosus** has an olive-brown downy cap, bright yellow tubes, and a yellowish-brown stalk, usually with slightly raised rusty-coloured veins. The pale lemon-yellow flesh can often be seen through cracks in the skin of the cap; sometimes it turns very faintly blue when broken.
B. impolitus is distinguished from *B. subtomentosus* by its unpleasant smell, and it is much less common.

1 Tylopilus Felleus   2 Boletus Piperatus   3 Boletus Edulis
4 Boletus Reticulatus   5 Boletus Erythropus
6 Boletus Chrysenteron   7 Boletus Subtomentosus

1 **Marasmius ramealis** is often found in large numbers in summer and autumn growing on twigs of various kinds, but especially bramble or, as shown here, wild rose. The white cap frequently has a flesh-pink tint, and both the distantly-spaced gills and the spores are also white. The surface of the stalk has a mealy texture. Some rather similar species of *Marasmius* are described on p. 46.

2 **Crepidotus variabilis** may be found throughout the year growing on sticks. The gills often face upwards, for the stalk is either very short and joined to the side of the cap, or it is lacking altogether, and the fungus is then attached by what would normally be its upper surface. The greyish-white cap is covered with very short hairs that give it a silky appearance; it has an incurved margin. The gills, white when young and rather distantly spaced, become a pale cinnamon colour with age. The spores are pale brown.

3 **Crepidotus mollis** grows on wood in summer and autumn. The cap is cream coloured when young, becoming white when dried, but it changes to a yellowish brown with age. It has a thick, jelly-like skin which gives the whole fungus a spongy texture. The gills are crowded and a pale cinnamon colour, but become darker and sometimes flecked with brown spots in older specimens. The very short stalk is joined to the side of the cap. The spores are pale brown.

Paxillus panuoides is somewhat similar to *Crepidotus mollis*, but the cap is dry, and the gills are easily separated from it. It is found on decaying stumps and branches of conifers, and sometimes attacks timber.

4 **Panellus stipticus** may be found throughout the year growing on stumps and fallen branches, often in overlapping clusters. The pale brown cap has a slightly mealy appearance; the gills are also pale brown and crowded, and the spores are white. The short stalk is joined to the side of the cap. The whole fungus has a leathery texture, although it becomes somewhat softened when wet, and it has a very bitter taste. When moist and actively growing, it sometimes glows in the dark, creating a striking effect.

Panellus mitis is similar to *P. stipticus*, but paler in colour, and only found on the wood of conifers, especially spruce. The flesh is spongy rather than leathery in texture and does not taste bitter.

5 **Mycena corticola** appears in late autumn on trunks and branches of various trees, often growing amongst mosses, and also, as shown here, on wood. The tiny grooved cap is very variable in colour, and is usually dark purple when young, but becoming bright blue, lilac, pink, or brown with age. It has a whitish bloom. The distantly-spaced gills are purple, paler than the cap, and becoming lighter still as the white spores are shed. The stalk has a thick bloom and is covered with short stiff hairs at the base. Other species of *Mycena* are illustrated on pp. 33, 105, 127, 132.

6 **Nectria cinnabarina** ('Coral Spot') is common at all times of the year on moist, newly fallen twigs and branches. It consists of small, dark-red, granular cushions, the colour of cinnabar, which burst through the bark, as shown on the right-hand part of the twig in the illustration. The spores are developed within very tiny flask-shaped cavities embedded in the cushions. Another even more common stage in the life cycle of this fungus, which is given the name *Tubercularia vulgaris*, consists of pink, powdery cushions, as shown on the left-hand part of the twig, the powder being a different kind of spore (coridia). A related species, *Dialonectria galligena*, causes canker in apple trees.

7 **Hypoxylon fragiforme** is found on dead trunks and branches, particularly of beech, in autumn and winter. The minute flask-shaped spore-producing structures develop just below the surface of almost spherical bodies, which are salmon pink when young, becoming red later, and finally very dark brown or almost black, as shown in the picture. The flesh is also dark brown or black and it is very brittle.

8 **Calycella citrina** grows on decaying wood in autumn and early winter. The spore-producing layers develop within shallow cups which are crowded together in large numbers. They are bright yellow but become orange yellow on drying. The outside of the cup and its very short thin stalk are a paler colour.

Helotium calyculus is similar to but less common than *Calycella citrina*. The cups are white on the outside and have longer, white stalks, which are downy at the base.

9 **Diatrype disciformis** appears in autumn and winter on dead branches of beech and other trees. The cushions, which contain the very small flasks in which the spores develop, burst through the bark, and are flat topped with an angular outline. They are pale buff at first, but become dark brown later and finally almost black, but the flesh remains white.

1 MARASMIUS RAMEALIS   2 CREPIDOTUS VARIABILIS   3 CREPIDOTUS MOLLIS
4 PANELLUS STIPTICUS   5 MYCENA CORTICOLA
6 NECTRIA CINNABARINA   7 HYPOXYLON FRAGIFORME   8 CALYCELLA CITRINA   9 DIATRYPE DISCIFORMIS

1 **Ustulina deusta** forms a thick, irregular crust on stumps and dead roots of beech and other trees, and may be found at all times of the year. The surface is greyish white when young, becoming black with age, and the flesh is white. The flask-shaped spore-producing cavities have conspicuous raised openings, from which come the black spores. In ripe specimens the flesh breaks down and the surface crust becomes very brittle, and then the fungus is easily crushed and detached from the wood on which it is growing.

2 **Ramaria stricta** grows on old stumps and rotting wood in the soil in late summer, autumn, and winter. It is pale yellow, often tinged brown, and has a short, thick stem with many slender branches which are clear yellow at the tips. The flesh is tough and becomes brownish grey or tinged with violet when bruised; it has a bitter or slightly peppery taste, and a faint, spicy smell not unlike aniseed. The spores are cinnamon coloured, not white as in *Clavaria* and its relatives (p. 153). Several other species grow on soil, but none of them is common.

3 **Coryne sarcoides** grows in groups on decaying stumps and old fallen logs of beech and other trees in autumn and early winter. It is reddish purple and jelly-like in texture. There is usually a short, thick stalk, on top of which is a small shallow cup with an irregular rim surrounding the spore-producing surface. The cup is very finely powdery on the outside and sometimes it has no stalk. Similarly coloured, somewhat irregular masses without cups, which produce a different kind of spore (conidia) may be found. In this state the fungus has been called '*Piriobasidium sarcoides*'.

4 **Bulgaria inquinans** forms large cups which have a soft elastic texture like rubber and grow in crowded clusters in autumn on the fallen trunks and branches of many trees. The upper spore-producing surface of each cup is black, smooth, and shiny, and the outer surface is sooty brown, covered with a thick coat of very small scales. The black spores are discharged in considerable quantity so that the plant is frequently surrounded by a conspicuous deposit of them. This helps to distinguish it from the superficially similar *Exidia glandulosa* (p. 149), which has colourless spores.

5 **Xylosphaera hypoxylon** ('Candle Snuff') has straight, black stems, usually somewhat flattened and with a hairy surface, which may be found at all times of the year growing on dead wood. The stems are forked at the tip, often more than once, and the flesh is white and very tough. Two kinds of spore are produced: at first, the tips of the branches become light grey with a powder of white spores (conidia); later they darken, and become covered with the raised openings of minute flask-shaped cavities within which black spores develop. Unbranched stems produce the black kind of spore only.

6 **Xylosphaera polymorpha** ('Dead Men's Fingers') is found at all times of the year growing on stumps of many trees, particularly of beech, often in clumps and usually at soil level. It is dull black, although the flesh within is white and solid, and it has a short, twisted stem with a large, irregularly-swollen upper part, which produces two kinds of spore. At first, the surface is smooth and becomes covered with a layer of light-brown spores (conidia); later, when the flask-shaped spore-producing cavities develop, it becomes rough to the touch as their minute raised openings protrude. These spores are black.

7 **Daldinia concentrica** ('King Alfred's Cakes', 'Cramp Balls') grows on dead trunks and large branches of ash, and very occasionally on other trees: small forms have been found on birch and gorse after heath fires. It appears and grows quickly in early autumn, and persists through the winter. The black spores are produced in very large numbers over a period of several months and are not shed until the following summer. The rounded fungus is dark reddish brown when young, but when fully grown it is black and shiny. The flesh is hard and purplish brown, with alternating darker and lighter zones. A remarkable feature of this plant is that when it is ripe it contains its own store of water and is able to discharge its spores however dry the weather; in fact it will continue to do so for some weeks even if it is detached from the tree and brought into a dry room. Normally the spores are shed only during darkness. The name 'Cramp Balls' refers to an old belief in its value as a charm against cramp and ague. Its presence in ash trees affects the timber, causing a condition known as 'calico wood'.

1 Ustulina Deusta

2 Ramaria Stricta   3 Coryne Sarcoides   4 Bulgaria Inquinans

5 Xylosphaera Hypoxylon   6 Xylosphaera Polymorpha   7 Daldinia Concentrica

147

The TREMELLALES, or 'Jelly Fungi', have a striking appearance because they are often brightly coloured when wet. They are usually very irregularly shaped and are softly jelly-like to the touch. In dry air they lose large quantities of water by evaporation, shrink very considerably, and when quite dry, become rigid and horny in texture and inconspicuous. When wetted, they swell up again and regain their colour. Over 500 species are known, but they are mostly tropical; about 50 species have been found in Britain.

Another group of jelly-like fungi are the Myxomycetes or 'Slime Fungi', (*see* p. 173). But these, when actively growing, have no fixed shape but consist of a mass of 'jelly' moving about slowly on the surface of rotting wood or other decaying plant material. Later, growth and movement cease, and large numbers of spore cases appear on short stalks.

1 **Auricularia auricula** ('Ear Fungus'), though to be seen at all times of the year, is especially common in autumn. In Britain and Europe it is almost entirely restricted to elder, but in North America it grows on a variety of trees. It is to be found, often in groups, on either living branches or dead wood, and varies from a liver brown to a dark flesh colour. The fungus is irregularly saucer-shaped, with inrolled margins and surfaces thrown into shallow irregular folds. The inner, spore-producing surface is shiny, while the outer surface is very finely velvety, with some greyish-olive tints. The flesh is slightly translucent, and is rather firmer in texture than most related species, feeling rather like very soft and flexible rubber. A related species, *A. polytricha*, is cultivated on oak palings in China for use as food.

2 **Auricularia mesenterica** is found at all times of the year on logs and on the trunks of a variety of trees. The spore-producing surface is brownish purple and much wrinkled, the folds forming an irregular network. The outer surface is brownish grey with zones of a darker colour, and has a markedly velvety appearance.

3 **Calocera viscosa** grows on the stumps of conifers in autumn and throughout the winter. It is bright orange yellow and its short stem forks repeatedly to form several long, straight branches. At a first glance, this fungus is rather like *Clavinulopsis corniculata* (p. 39), but it is jelly-like in texture, and the surface is slimy to the touch.
C. cornea is a smaller and paler species than *C. viscosa* and grows in tufts. *C. stricta* is similar but is yellow, grows singly, and is never branched. *C. striata* differs from *C. stricta* in than it becomes wrinkled on the surface when dry.

4 **Exidia glandulosa** ('Witches' Butter') appears on dead branches, especially of oak, most frequently in winter, but also at other times of the year. The spore-producing surface is covered with small raised warts and is very dark brown when young, and the outer surface is dark grey and slightly downy. The whole plant soon turns entirely black. It is not unlike *Bulgaria inquinans* (p. 147), but the latter has a smooth surface and black instead of colourless spores.

Exidia recisa is circular in outline, about the size of a halfpenny, slightly translucent, and yellowish brown in colour. It grows attached by a short stalk to willow branches. There is a similar species *E. truncata*, which is black and usually found on oak.

Exidia albida forms white, irregular masses which burst through the bark of decaying branches. It becomes greyish brown with age.

5 **Tremella mesenterica** may be seen at all times of the year growing on dead branches, but is especially common in late autumn and early winter. It forms a clear orange-yellow, rather firm jelly-like mass which is thrown into wavy folds.

6 **Tremella foliacea** grows on the stumps of pine and other trees at all times of the year. It consists of a jelly-like mass folded irregularly into a large number of thin plates attached in tufts to a wrinkled base. It is pinkish cinnamon in colour, often with a violet tinge.

7 **Dacrymyces deliquescens** is to be found at all times of the year on wet, rotten wood. It consists of small, orange, jelly-like cushions, often growing in lines from cracks in the wood. A very similar species, *D. stillatus*, is a deep orange colour and has a rather firmer texture.

1 Auricularia Auricula   2 Auricularia Mesenterica   3 Calocera Viscosa
4 Exidia Glandulosa
5 Tremella Mesenterica   6 Tremella Foliacea   7 Dacrymyces Deliquescens

149

The PEZIZALES or Cup Fungi have a spore-producing layer which develops within a more-or-less shallow cup. Since this faces upwards, the spores when ripe cannot fall out, and so are distributed by being discharged upwards with considerable violence. If a ripe cup is tapped sharply, the spores will appear as a cloud looking like a thin mist above the surface. In some of the larger species, for example *Otidea onotica*, a mere touch or breath upon the fungus may be sufficient to stimulate spore discharge. These are mainly plants of the northern hemisphere, and over 500 species are known, about 250 of which have been found in Britain. The majority have cups considerably smaller than a sixpence, and many are smaller than the head of a match; so only a few of the commoner larger ones are described here. *Morchella esculenta* (p. 41) and *Helvella lacunosa* (p. 153) are regarded as members of the Pezizales, although their spore-producing structures are not simply cup-shaped.

1 **Humaria hemisphaerica** grows on the soil in woods, or occasionally on very rotten wood in wet places, in summer and autumn. The outside of the cup is brown and entirely covered with very dark-brown hairs; the inside is white or very pale brown.

2 **Aleuria aurantia** ('Orange Peel') is found in autumn and winter on bare ground in woods and in the open, often by footpaths. The cups grow in close groups and press on each other so much that they often become irregular in outline and split at the margins. They are bright orange on the inside, and the paler outside is covered with a soft white down.

3 **Peziza repanda** grows on the ground, usually associated with wood fragments, and is often found on old sawdust heaps. The cups appear in the autumn and are a light chestnut brown on the inside and pale fawn on the outside, which is covered with a thin coating of very small scales; the flesh is also fawn coloured. *P. varia* is similar to *P. repanda*, but the inside of the cup is light greyish brown, becoming darker as, with age, the cup flattens and finally turns under at the margins. Though commonest in summer, it is also found in the autumn. *P. badia*, which is also found in woods in summer and autumn, is olive-brown inside and reddish-brown outside. The flesh is a pale reddish brown with a watery juice. *P. anthracophila*, which grows on burnt soil and charcoal from late spring until autumn, has a cup with a dark-brown inside and brown flesh. The margins of the older specimens are strongly incurved.

Peziza vesiculosa may be found at any time from late summer until the following spring growing on manure heaps or richly manured soil. The outside of the cup is pale fawn, covered with rather coarse scales, and the very brittle flesh is also fawn. The spore-producing layer often becomes detached from the inner surface of the cup to form raised blisters.

4 **Sarcoscypha coccinea,** which is particularly common in the West Country, grows on decaying branches on damp ground in woodlands. It is most abundant in winter, but is also found in early spring. The cups grow in groups; the inside is scarlet, and the outside white, covered with a thick coat of matted hairs.

5 **Otidea onotica** ('Lemon Peel') is found in early autumn growing on the ground, especially in oak woods. The scoop-shaped 'cups' usually grow in groups, and are yellow, flushed with pink on the inside, and tinged pale brown with a finely mealy texture on the outside. There is a similar but rarer species, *O. concinna*, which is smaller and a bright sulphur-yellow colour. Other uncommon species of mixed woodlands are *O. alutacea*, which is greyish brown on the inside, and *O. umbrina*, which is dark brown inside and pale brown outside.

Sarcosphaera eximia is found especially in beech woods on chalky soils in the spring. At first it is a hollow, smooth, white ball, almost hidden in the ground, which may grow as large as a goose's egg. When ripe, it splits open to expose the violet-coloured spore-producing surface.

6 **Chlorociboria aeruginascens** grows on rotting logs and branches, especially of oak, staining the wood a bright blue green. The cups, which usually appear in the autumn, though sometimes in spring and summer, are also blue green. Snuff boxes, jewel caskets, and such small wooden objects, inlayed with this green wood, used to be manufactured at Tonbridge in Kent until late in the 19th century and became known as 'Tonbridge Ware'. This fungus has been called '*Chlorosplenium aeruginosum*', but the name properly belongs to another species not found in Britain, in which the inside of the cup becomes yellow on drying.

1 HUMARIA HEMISPHAERICA   2 ALEURIA AURANTIA
3 PEZIZA REPANDA   4 SARCOSCYPHA COCCINEA   5 OTIDEA ONOTICA
6 CHLOROCIBORIA AERUGINASCENS

1 **Phallus impudicus** ('Stinkhorn','Wood Witch') appears in rich soil in woods and gardens, from midsummer until late in the autumn as a white oval body pushing through the surface of the ground, and grows until it is about the size of a goose's egg. If it be cut open vertically at this stage, the structure of the fully-grown fungus can be seen. Within the white outer skin (the 'universal veil') is a conical cap covered with a thick layer of olive-green jelly containing the spores, and a white spindle-shaped stalk pointed at both ends. When the 'egg' is ripe, the stalk elongates very greatly in the course of a few hours, tearing the veil and carrying the cap upwards. It is a process of expansion rather than growth, the stalk changing from a compact structure to a hollow porous one. Margaret Plues, in her book *Flowerless Plants* (1865), remarks that 'The plant has a dignified and imposing appearance, and might well be accounted a desirable ornament of woods and pleasure grounds, but for its all-pervading odour'. A rapid breakdown on exposure to air of substances in the jelly does indeed produce a powerfully foetid stench. This attracts flies, which feed on the slime, and thus distribute the spores.

2 **Mutinus caninus** ('Dog Stinkhorn') differs from *Phallus impudicus* in its smaller size, its colour (the 'egg' being yellowish white, the stalk yellowish brown, and the cap, beneath the dark green spore mass, orange red), and its considerably less powerful odour. It is to be found, rather less commonly, in woodlands from midsummer until late autumn.

3 **Hydnum repandum** ('Wood Hedgehog') is found in woodlands in late summer and autumn, often growing in groups or rings. The cap is a very pale yellowish brown, sometimes tinged with pink and covered on the underside with a layer of spines or 'teeth'. The specimen illustrated is a particularly robust one. A smaller species, *H. rufescens*, has a reddish-brown cap and spines. Both species have white or whitish-yellow flesh, which is rather tough and has an acrid taste. They are edible if well boiled first.

Hericium erinaceum is a rare plant consisting of a yellowish-white, irregularly-lobed, and shortly-branched fleshy mass covered with hanging, pointed spines. Apparently it is properly called 'Hedgehog Fungus', but because of its rarity the name has become transferred to the related species *Hydnum repandum*.

4 **Helvella lacunosa** appears in autumn in woodlands, often on burnt soil. The dark grey cap consists of two lobes which arch away from the stalk, and are turned under at the margins. The grey and deeply-grooved stalk has one or more hollow chambers within. The flesh is dark grey, sometimes with a lavender tint.

Helvella crispa grows in woodlands, especially beside paths, in late summer and autumn, and also in early spring. The cap is pale whitish grey, and the lobes are irregularly folded and wavy at the margins. The stalk is white, deeply grooved, and hollow. The flesh is white or cream coloured.

Gyromitra esculenta is an uncommon plant of conifer woods on sandy soils, most often found in Scotland. The chestnut-brown cap has a very wavy and wrinkled surface, and the whitish-grey stalk is only slightly grooved. It is poisonous when raw, though it has been eaten after being thoroughly boiled and the water thrown away. However, this fungus and *Helvella lacunosa* and *H. crispa*, which are sometimes called 'False Morels', should not be eaten.

5 **Clavulina cristata** is found in the latter half of the year growing in woodlands, especially beside paths. It has many short white branches, which are sometimes tinged with pink or grey, and the flesh is brittle. This fungus has been called *Clavaria cristata*, but there are important microscopical differences (for instance, the spores develop in pairs rather than in fours). A larger species, *Clavulina rugosa*, is illustrated on p. 39.

Clavaria argillacea grows in open places amongst grasses and mosses, frequently on peaty ground, in summer and autumn. It is club-shaped, unbranched, whitish yellow at the base, and yellowish buff at the tip, and the flesh is brittle.

6 **Clavulinopsis helvola.** An autumn and winter woodland fungus, also found in the open; it is orange yellow, unbranched, and tapered towards the tip. The flesh is firm and waxy. A similar species, *C. vernalis*, about one third as tall and a dull brownish-orange colour, has never been found in Britain, though it does grow in France and the Netherlands. It is interesting because it always grows on soil that is covered with a layer of the one-celled blue-green alga *Gloeocapsa*. A bright-yellow and much branched species, *C. corniculata*, is illustrated on p. 39.

Clavariadelphus pistillaris is yellowish-brown with spongy flesh, and is shaped like a pestle. It grows in woodlands in autumn and winter.

1 PHALLUS IMPUDICUS   2 MUTINUS CANINUS   3 HYDNUM REPANDUM
4 HELVELLA LACUNOSA   5 CLAVULINA CRISTATA   6 CLAVULINOPSIS HELVOLA

In the GASTEROMYCETES the spores are formed inside the fungus, and not by an exposed surface layer; consequently their methods of dispersal are more varied and complex. In *Lycoperdon perlatum*, for instance, the slightest blow on the ripe fungus causes the spores to puff out in a cloud; and since they are not readily wettable, rain drops falling amongst them will cause puffing also. In *Phallus impudicus* (p. 152) the spores are dispersed by flies; and in *Melanogaster variegatus*, which grows underground, they are distributed by small animals. *Tuber aestivum*, which is not a member of this group, is included here because it is similarly distributed.

1 **Geastrum triplex** ('Earth Star') is found in summer and autumn, often in beech woods, but also associated with other kinds of trees. It is pale brown, often tinged faintly pink or almost flesh colour, and shaped like an onion when young. It becomes darker with age. The wall has three layers. As it develops, the outer layer splits into about six lobes which fold back, forming a star-shaped pattern on the ground. The inner layer contains the purplish-brown spores, which eventually escape through a small raised opening at the top, which is fringed with silky hairs. The middle layer is fleshy, and forms a shallow cup holding the spore-containing part. Ten other species of *Geastrum* have been found in Britain, but none of them is common.

2 **Lycoperdon perlatum** ('Puff Ball') is a woodlands species of summer and autumn and often grows in groups. The whole plant is whitish grey when young, becoming yellowish brown with age. The upper part, which when fully developed is completely filled with a mass of olive-brown spores, is almost spherical and is covered with intermingled long spines; these readily fall off, leaving pale spots and short pointed warts. The surface of the thick stalk is covered with smaller warts which are not pointed. When the spores are ripe, a raised pore (rather like a miniature volcano) appears at the top, through which they can escape. *L. molle* (sometimes called *L. umbrinum*) is similar but has a much shorter stalk, usually about one-third the height of the whole fungus; it is most frequently found in conifer woods. The rather rare *L. echinatum* has no stalk and is covered with crowded long brown spines.

3 **Lycoperdon pyriforme** appears in late summer and autumn, growing on old stumps or the dead roots of trees. It is the only species that grows on wood. It is pear-shaped, white, and covered with small thick scales that give it a mealy appearance when young; it becomes brownish grey and smooth with age. The plant has a not very powerful but distinct smell, rather like that of a fresh herring.

4 **Scleroderma verrucosum** grows in woodlands on rich soil in summer and autumn. It may be almost spherical or, more frequently, bun-shaped, and has hardly any stalk. When ripe, it is completely filled with a mass of very dark olive-brown spores. The wall is pale greyish brown, thin, and covered with small scales when young, but becomes smooth with age. Eventually it cracks open in an irregular manner, and the spores are released.

5 **Scleroderma aurantium** ('Earth Ball') may be found in summer, autumn, and early winter and is frequently but not exclusively associated with birch trees. It is similar in shape to *S. verrucosum*, but has a thick, tough, brownish-yellow wall, covered with flat brown warts. When ripe, the purplish-black spores are released through irregular cracks in the wall. The fungus has a somewhat sour smell, rather like that of ink.

Melanogaster variegatus is related to *Scleroderma* but develops entirely underground. It resembles a small reddish-brown potato in shape and size. Inside it is divided into chambers in which the brown spores are embedded in a purplish-black jelly. The variety *broomianus* used to be sold for food in the West Country under the name of 'Red Truffle'.

Tuber aestivum, like *M. variegatus*, develops underground but has a quite different microscopical structure and is, in fact, more closely related to *Morchella esculenta* (p. 41). The skin is dark brown and covered with prominent warts; inside it is greyish buff, often tinged violet, with a network of white veins. It has a nutty, slightly earthy, flavour. *T. melanosporum* (Perigord Truffle) does not grow in Britain but is common in parts of France.

Elaphomyces granulatus ('False Truffle'), which is found mainly in conifer woods, is rather distantly related to *Tuber*, and also grows entirely underground. It looks rather like a small specimen of *Scleroderma aurantium*, but the wall surrounding the dark blackish-brown spores is distinctly two-layered.

6 **Boletus parasiticus** differs from other species of Boletus (*see* p. 142) in that it is parasitic on another fungus. It may be found in autumn growing on *Scleroderma aurantium* or *S. verrucosum*. The cap is brownish yellow with an olive tinge; the tubes are golden yellow; and the stalk is yellow, often streaked with red. A curious feature of this fungus is that it seems hardly ever to release its spores.

1 Geastrum Triplex   2 Lycoperdon Perlatum   3 Lycoperdon Pyriforme
4 Scleroderma Verrucosum   5 Scleroderma Aurantium   6 Boletus Parasiticus

1 **Stereum purpureum** is found at all times of the year on the living branches of plum, blackthorn, and related trees, and also sometimes on dead timber from the same kinds of tree. It forms crusts or sheets, which are hard to the touch and have tough, only slightly flexible flesh; they partly lie flat on the bark and partly project horizontally, often forming several tiers of small brackets. The spore-producing layer, which is lilac coloured when young, becoming grey with age, is on the underside of the projecting parts and is exposed on the visible surface of the basal parts. The upper sides of the brackets are dull greyish brown and have a velvety texture. The fungus causes 'Silver Leaf' disease of plums and occasionally of other fruit trees. Whether or not an attack will be serious seems to depend on the general health of the tree, and vigorously-growing trees sometimes recover. Over-drastic pruning by weakening the tree may encourage the spread of the disease. The varieties 'Victoria' plum and 'Newton Wonder' apple are very susceptible, while others are seldom affected.

2 **Stereum hirsutum** grows throughout the year on logs, fence posts, and standing and fallen timber of all kinds. It is bracket-shaped, and often grows in tiered clusters. The under, spore-producing surface is a bright yellow (or sometimes orange) brown; the upper surface is a dull yellowish grey, and is covered with felted hairs.

3 **Stereum rugosum** is found on stumps and fallen trunks and branches, and also on living trees, at all times of the year. It usually lies flat on the wood and is firmly attached, with the yellowish-buff spore-producing surface upwards; this becomes grey with age and stains blood red when cut or bruised. The margins are upturned, and the other side is brown and wrinkled.

Stereum sanguinolentum resembles *S. rugosum* in that it usually lies flat and stains red when bruised, but it is thinner and is only found on the dead wood of coniferous trees. Other related and somewhat similar species are described on p. 116.

4 **Gloeoporus adustus** grows in groups on trunks, stumps, and fallen trees of all kinds, though rarely on conifers, at all times of the year. It forms crowded tiers of brackets, the upper surfaces of which are pale brownish grey, with indistinct darker zones, and are closely covered with very short hairs. The margins and lower surfaces are greyish white when young, but darken with age, finally becoming black. The greyish-white flesh is thin, tough, and flexible, and is separated from the layer of tubes on the underside by a thin, grey or almost black, jelly-like layer, which appears as a dark line when the fungus is cut or broken. The tubes have extremely minute openings. This fungus causes white flecks in the sap wood of felled trees. A similar but usually larger species, *G. fumosus*, is described on p. 125.

5 **Phellinus pomaceus** is found at all times of the year, especially on plum and blackthorn, but also on hawthorn and related trees. It forms a very hard rounded bracket which is difficult to detach from the branch or trunk of the tree. The upper surface is brownish-grey and smooth and has a fringe of very short hairs on the margin only; in old specimens the hairs disappear, and the surface may become finely cracked. The tubes on the lower surface are brown with minute openings which only appear after a long period of development. *P. pomaceus* is frequently perennial, producing a new layer of pores each year. It causes an extensive heart rot in living trees, the wood becoming soft and stringy and stained pale brown, with darker flecks and streaks.

Phellinus ignarius is usually larger than *P. pomaceus* and is more prominently bracket-shaped with a flatter, dark greyish-brown or black upper surface, which is deeply furrowed and cracked. The tubes are whitish grey when young, becoming cinnamon coloured with age. It grows on willows and poplars, and is sometimes found on birch, when it could be confused with *Fomes fomentarius* (p. 117).

6 **Inonotus radiatus** is found on alder and birch, and occasionally on other trees, usually on dead trunks, at any time from autumn until the following spring. It forms groups of brackets which may overlap one another. The upper surface is a bright yellowish brown, often with an orange or greenish tinge, when young, and becomes rusty brown with age; it is radially furrowed. The tubes are pale greyish brown and have minute openings. This fungus stains the wood on which it is growing a rusty yellow colour.

Inonotus hispidus grows on various trees but especially ash. The rusty-brown upper surface is covered with a dense coat of shaggy hairs. The tubes are yellowish brown, often with an olive tinge, and have minute openings.

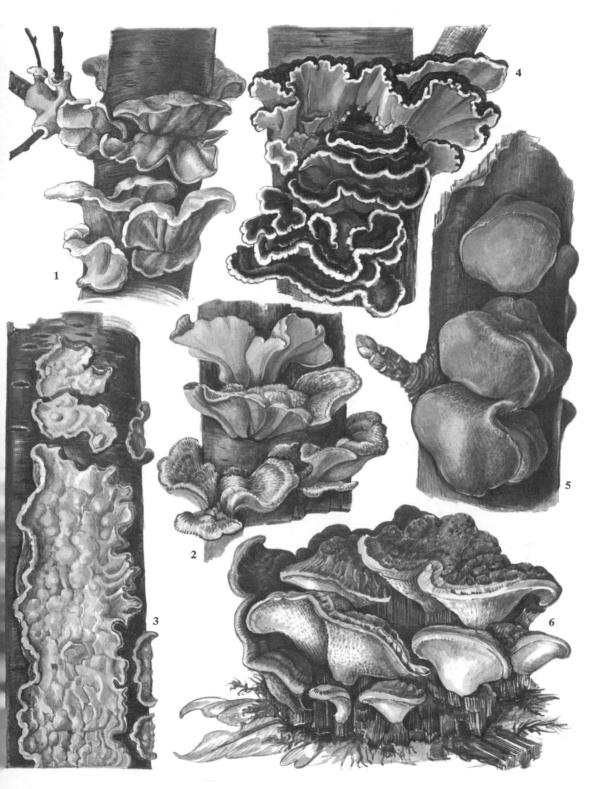

1 STEREUM PURPUREUM   2 STEREUM HIRSUTUM   3 STEREUM RUGOSUM
4 GLOEOPORUS ADUSTUS   5 PHELLINUS POMACEUS   6 INONOTUS RADIATUS

The amount of light available to flowerless plants growing in woodlands is considerably diminished by the shade cast by the trees. Fungi, which can grow in darkness, are the least affected by this, although many of them turn their spore-producing parts towards the light. Woodland ferns, mosses, and liverworts are shade tolerators, although the latter two groups grow most actively in the spring before the canopy of foliage is complete. Lichens are, on the whole, light demanders and grow most actively when the trees are not in leaf, although many species, for instance *Graphis elegans* (p. 171), appear to require a mixture of some direct sunlight and some shade.

Nearly forty species of the genus *Usnea* have been found in Britain. They all have long stems growing from small disc-shaped holdfasts, and they branch repeatedly to form a tangle of hair-like threads. When cut across, the stems and branches show a central core which is tough and elastic, an outer skin (cortex), and between these a soft, usually white, layer (medulla). If a short length of the stem is pulled between the fingers, the core stretches, but the cortex breaks, revealing the medulla within.

1 **Alectoria fuscescens** is a plant which is common in the north but much less so in the south, and grows on trees and fences, and sometimes on rocks. The thin stems grow from small holdfasts, and divide repeatedly to form smooth, shiny, thread-like branches, which vary in colour from greyish brown to brownish black. There is no central core in the branches, such as is found in *Usnea* species. Usually there are scattered small patches of pale-yellow or green reproductive granules (soredia) on the branches.

2 **Usnea articulata** grows on trees, especially in the West Country. It hangs downwards, forming long festoons of branches, which are soft, smooth, and shining grey green; the older stems become swollen, with constrictions at intervals, giving the appearance almost of lengths of intestine or irregular strings of sausages.

3 **Usnea ceratina** is common on trees in England, though less so in Scotland and Ireland. Young plants grow erect, but come to hang downwards with increasing age and size. The bluish grey-green branches are rather stiff and somewhat angular in cross sections, and there are numerous white floury patches on the surface (see enlarged detail 3A). The powder consists of minute reproductive structures (soredia). There are also very small wart-like outgrowths, and short, smooth cylindrical ones as well. The medulla or inside layer has a rose-pink tint.

Usnea filipendula, common in the north of Scotland but rare in England, differs from *U. ceratina* in that it never grows erect, and there are rod-like cylindrical structures (isidia) amongst the reproductive soredia.

4 **Usnea rubiginea** is one of the commoner species in woodlands, especially in England and Wales. It grows

erect, though some branches in old plants may hang downwards, and it is always red, at least in parts. There are white patches on the branches like those of *U. subfloridans* (5A), and small wart-like outgrowths also.

5 **Usnea subfloridans** is a common species in woodlands and may sometimes be found growing on rocks as well as on trees. It grows erect, though some of the branches in old plants may hang downwards, and is dark greenish grey. There are white patches on the branches (5A) which produce minute reproductive structures in the form of a powder (soredia) mixed with short rods (isidia), and short, smooth cylindrical outgrowths also.

Usnea flammea and **Usnea subpectinata** are the only other species that grow both on trees and on rocks, but they are much less common than *U. subfloridans*. *U. flammea* has no cylindrical outgrowths on the branches, and *U. subpectinata* has coarser soredia which are like fine sugar rather than like flour.

6 **Usnea florida** grows on trees, frequently in the open or in hedgerows, and is common, especially in the West Country. It grows erect, and is dark greenish grey. Its large cup-shaped spore-producing structures (apothecia), rare in most other species of *Usnea*, grow on the ends of the branches; their upper surfaces are pale yellow, and there are thick long hairs radiating from their margins.

7 **Teloschistes flavicans** is found on trees and rocks, often near the sea, especially in the West Country. It forms upright or drooping slender tufts of slightly flattened slender branches, which are golden yellow, sometimes tinged with orange.

1 ALECTORIA FUSCESCENS   2 USNEA ARTICULATA   3 USNEA CERATINA
4 USNEA RUBIGINEA   5 USNEA SUBFLORIDANS   6 USNEA FLORIDA
7 TELOSCHISTES FLAVICANS

1 **Ramalina calicaris** grows on the trunks and branches of trees throughout Britain, although less commonly in the north; it has been found attached to lightly buried twigs on sand dunes. The flattened branches grow in erect tufts and are a shiny, pale greenish grey in colour. A shallow channel runs along their length, and the surface is often irregularly pitted or dotted with small circular openings (pseudocyphellae). The spore-producing structures (apothecia) are pale flesh-coloured discs, often with a greenish tinge, growing from the tips and the margins on short stalks, which are frequently bent at an angle.

Ramalina fastigiata is a common species, similar to *R. calicaris*, but the rather short branches are not channelled and are only sparingly branched, so that the plant has a densely tufted appearance. The spore-producing apothecia are disc-shaped structures at the tips of the branches and are usually numerous.

2 **Ramalina fraxinea** grows on tree trunks, and makes especially well-developed plants in the West Country and in west Scotland, where the branches may develop into broad ribbons as long as, or longer than, a man's index finger. The rather dark grey-green surface is covered with a network of wrinkles, and is perforated by circular or elongated openings. The spore-producing apothecia grow on short stalks from the margins, and also from the surface.

3 **Ramalina farinacea** forms dense tufts of narrow greenish-grey branches, and is common on the trunks and twigs of trees and shrubs. The margins have numerous small circular or oval patches of powdery reproductive bodies (soredia), looking like flour, but the disc-like apothecia are rare.

4 **Evernia prunastri** ('*Mousse de Chêne*') is a common plant on trees and fences and has flattened branches which fork several times. It may be distinguished from species of *Ramalina* because, although the upper surface is greenish grey, the under surface is white, lacking all trace of green. The upper side is covered with a network of wrinkles, along which powdery

soredia often develop. The disc-like apothecia are not very often found. The plant is used by perfumiers as a fixative for other ingredients in their recipes; in Poland the plant is protected by law and may only be gathered from felled trees.

5 **Evernia furfuracea**, sometimes called *Parmelia furfuracea*, grows on rocks and stone walls as well as on trees and fences. The under surface is dull black or sometimes mottled black and grey, and the dark-grey upper surface is often rough with small, almost bristle-like projections (isidia).

6 **Parmelia acetabulum** grows on tree trunks and has broad, dark brownish-green lobes, which have a somewhat oily appearance when wet and are dull and wrinkled when dry. Most lichens are commoner in the West Country than elsewhere, but this plant is almost entirely confined to the eastern parts of Britain. It is also found on the continent, especially in Central Europe, but is not known from other parts of the world.

7 **Parmelia exasperatula** is a dark-brown or yellowish-brown, shiny lichen growing on twigs, very closely attached to the bark. It is thickly covered with small warts (7a) which are unique in having tiny openings at the top; these have been called 'breathing pores' though their actual function is not known.

8 **Parmelia subaurifera**, like *P. exasperatula*, grows very closely attached to the bark of the smaller twigs of trees and shrubs. It is yellowish brown with a dull, never shiny, surface. Frequently it has small wart-like outgrowths (isidia) clustered on the surface (8a), but these have no aperture at the top, although sometimes yellowish-white, powdery reproductive bodies (soredia) develop at the tips.

Parmelia glabratula is similar to *P. subaurifera* and also grows on the twigs of trees. The surface of the plant is shiny, however, and not dull in texture.
Other species of *Parmelia* are illustrated on pp. 63, 163, and 165.

1 Ramalina Calicaris   2 Ramalina Fraxinea   3 Ramalina Farinacea
4 Evernia Prunastri   5 Evernia Furfuracea   6 Parmelia Acetabulum
7 Parmelia Exasperatula   8 Parmelia Subaurifera

For a general note on lichens see page 66.

1 **Parmelia saxatilis** (Crottle) is very common throughout Britain growing on trees and also on walls and rocks, everywhere except at high altitudes, where it is replaced by *P. omphalodes* (p. 63). It forms flat, grey rosettes of rather narrow lobes, which are slightly broadened and fan shaped at the ends. The surface is covered with a fine network of white lines, and numerous short, greyish-brown, rod-like outgrowths (isidia) develop in the centres of older plants.

2 **Parmelia sulcata,** which is similar to *P. saxatilis*, is even more common on trees, especially in southern and eastern England. Instead of rod-like outgrowths (isidia), it has white, powdery, reproductive structures (soredia) which develop along the network of lines on the surface. Other species of *Parmelia* are shown on pp. 63 and 164.

3 **Parmeliopsis ambigua** forms small yellowish-grey rosettes, usually near the bases of tree trunks, and has scattered, yellow, powdery reproductive structures (soredia) on the surface. Although it looks rather like *Parmelia mougeotii* (p. 62), which grows only on rocks, the lobes of *Parmeliopsis ambigua* are longer and narrower than in most species of *Parmelia*.

4 **Parmeliella plumbea** grows on trees and sometimes on rocks in the west and north of Britain. It forms rather soft, round patches, folded and ridged towards the margins and granular in the centre. These are bluish green when wet, but become bluish grey on drying. The spore-producing structures (apothecia) have reddish-brown discs with paler margins. The algal cells in this lichen belong to a species of the blue-green alga *Nostoc* (*see* p. 66).
P. atlantica is similar in appearance and grows in the same kinds of places, but the centre of the plant becomes covered with short rod-like outgrowths (isidia).

5 **Hypogymnia physodes** is very common on trees, rocks, and walls throughout Britain. It often grows with *Parmelia sulcata* and *P. saxatilis* and is a similar pale grey colour, but the upper surface is quite smooth. It differs from all *Parmelia* species in having an air space between its upper and lower surfaces, which gives it an inflated appearance. Also, instead of being secured by thread-like outgrowths (rhizines), it is attached to bark or stone directly by its lower surface and so cannot be detached without tearing. The under surface is black in the centre and brown at the margins. The spore-producing structures (apothecia), which are not commonly found, have brownish-red discs and grey margins. Greyish-white, powdery, reproductive structures (soredia) develop under the lobe ends. Another form of this plant is shown on p. 53.

6 **Hypogymnia tubulosa** is found most commonly in the north and west of Britain. It grows mainly on twigs and the trunks of trees, attached by its very dark-brown or black undersurface, and it has an inflated appearance. The lobes are tubular and grow upwards, with ball-like masses of greyish-white, powdery, reproductive structures (soredia) at their tips.

7 **Cetraria glauca** is common on trees, fences, walls, and rocks throughout Britain. It has broad, thin lobes, attached at the centre by threads (rhizines) but quite free at the margins, and with a wavy, crisped appearance. The upper surface is grey, often with a bluish tinge, and the under surface is brown, smooth, and shining, becoming black in the centre. The margins are often fringed with short, rod-like outgrowths (isidia) and sometimes with greyish-white, powdery, reproductive structures (soredia) also. A variety *fallax* has a completely white underside and is quite common in Scotland.

8 **Cetraria chlorophylla** grows in the same kinds of places as *C. glauca*, but is less common. It is a smaller plant, very much crisped at the margins, and brown on both surfaces. White, powdery, reproductive structures (soredia) develop along the edges. Two other species of *Cetraria* are shown on p. 67.

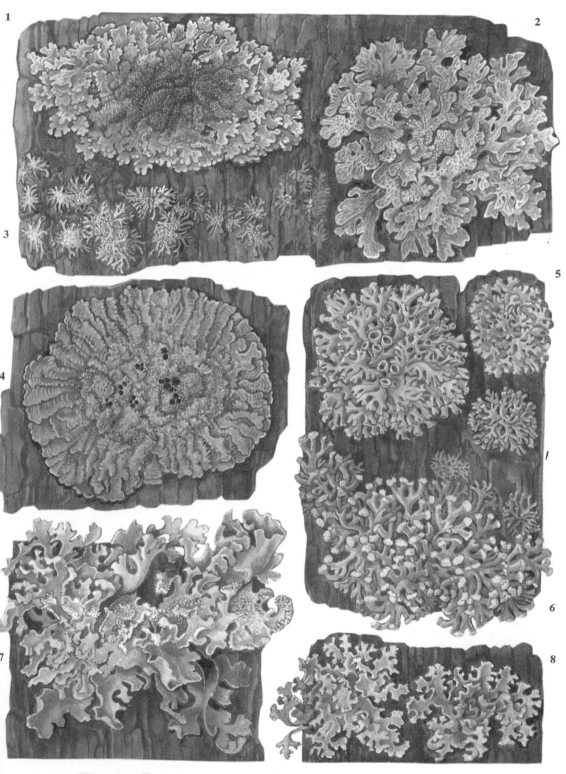

1 PARMELIA SAXATILIS   2 PARMELIA SULCATA   3 PARMELIOPSIS AMBIGUA
4 PARMELIELLA PLUMBEA   5 HYPOGYMNIA PHYSODES   6 HYPOGYMNIA TUBULOSA
7 CETRARIA GLAUCA   8 CETRARIA CHLOROPHYLLA

For a general note on lichens see page 66.

1 **Physcia leptalea** is common on the branches and trunks of trees, and is occasionally found on rocks. The grey, much-branched lobes are speckled with minute white dots (pseudocyphellae) and are rather loosely attached by scanty, thin threads (rhizines) on the underside. The edges are sparsely fringed with pale-brown, rather long, hairs, and the spore-producing structures (apothecia) have reddish-brown discs with grey margins. At a first glance, *Physcia* species somewhat resemble *Parmelia* species, but their upper surfaces are soft and velvety rather than hard and shining, and underneath they are dull and pale, especially at the edges, instead of shiny brown or black. The rhizines are fewer and fibrous, rather than thick and wiry, as in *Parmelia* species.

2 **Physcia pulverulenta** forms rosettes varying in colour from silvery grey to greenish brown. It grows closely attached to the bark of trees, or more rarely to rocks, by short, fibrous threads (rhizines). The rather broad, blunt-tipped lobes with slightly turned-up margins are dusted with a very fine powder (pruina). The spore-producing apothecia have prominent margins, and the discs are covered with pruina when young (2A) but become very dark brown or almost black when mature. A variety *venusta*, in which the margins of the apothecia develop leafy lobes and which is lacking in pruina, is quite common.

3 **Physcia aipolia** grows firmly attached to the bark of trees, forming rosettes of rather narrow, branching lobes. The surface is sprinkled with minute white dots (pseudocyphellae), and the spore-producing apothecia have very dark-brown or black discs, which are covered with a very fine powder (pruina) when young (3A).
P. stellaris is a similar, but much rarer, plant with whitish-grey lobes and no, or only a few, very indistinct pseudocyphellae. Other species of *Physcia* are described on pp. 76 and 168.

4 **Anaptychia ciliaris** is found on trees and occasionally on rocks or on the ground. It has long, narrow, branching, brownish-grey lobes which become greenish-grey when wet, and are fringed with long hairs at the margins. The spore-producing apothecia are large and have dark-brown or almost black discs, with elegantly fringed rims. A related species, *A. fusca*, is shown on p. 3.

5 **Parmelia subrudecta** is common on trees in England but rather rare in Scotland. It has rounded lobes and clusters of white, powdery, reproductive structures (soredia), which develop as white dots on the upper surface. Sometimes the margins are turned upwards, exposing the pale-brown surface underneath, as in the detail drawing 5A, which also shows some small lobes (folioles) and a cluster of soredia growing on top.

6 **Parmelia perlata** grows on trees and rocks, especially in the west of Britain, and forms large, light-grey patches, with rounded lobes which often have black hairs at the edges. Clusters of greyish-white, powdery, reproductive structures (soredia) grow on the up-turned tips of some of the lobes, and large spore-producing apothecia with dark-brown discs and prominent margins sometimes develop in the centre.
P. crinita is another common species in western districts and resembles *P. perlata*, though the margins are more finely lobed, and rod-like outgrowths (isidia), usually with clusters of soredia at their tips, develop on the surface.

7 **Parmelia caperata** is quite common, especially in southern England, where on old trees it often reaches a very large size. It also sometimes grows on rocks. The broad lobes are a beautiful yellow-green colour, not paralleled by any other plant. Coarse, yellow-green, powdery, reproductive structures (soredia) and large spore-producing apothecia with reddish-brown discs and prominent margins develop in the centre.

8 **Parmelia laevigata** forms large, light-grey patches on trees in the south and west of Britain. The lobes are forked, with large, round gaps between the branches, and the tips frequently have a 'cut off' appearance, and may develop globose masses of white, powdery, reproductive structures (soredia).
P. revoluta is a similar species, growing in the same kinds of places, but the lobes are forked at narrower angles, the tips rounded, the edges turned under, and the soredia scattered along the margins. Other species of *Parmelia* are shown on pp. 63 and 163.

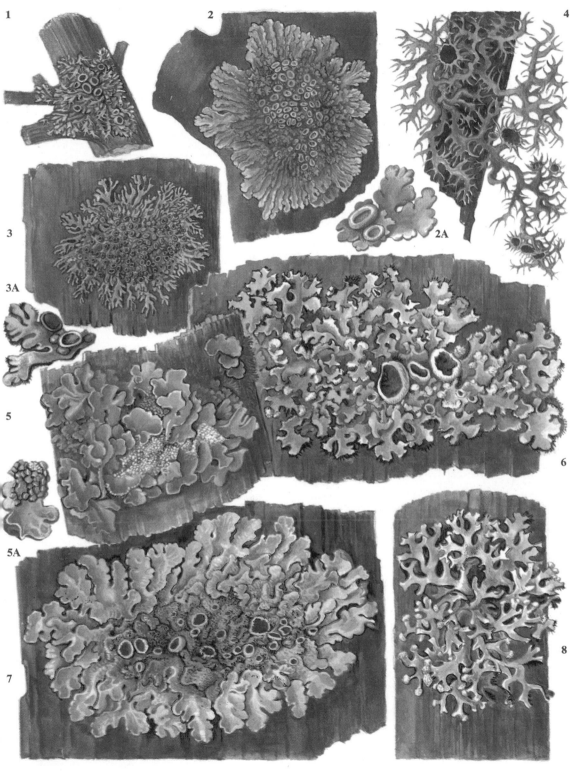

1 Physcia Leptalea   2 Physcia Pulverulenta   3 Physcia Aipolia
4 Anaptychia Ciliaris   5 Parmelia Subrudecta
6 Parmelia Perlata   7 Parmelia Caperata   8 Parmelia Laevigata

For a general note on lichens see page 66.

1 **Sticta limbata** grows amongst mosses on the trunks of old trees, or occasionally on rocks, in the western, wetter parts of Britain. The light-grey lobes are rather smooth and shining and develop a bluish tinge when wet. The margins and the upper surface are sprinkled with bluish-grey, powdery, reproductive structures (soredia), and the pale-brown under surface (1A) is perforated by neat round holes with distinct rims (cyphellae). *Sticta* species are very similar to *Pseudocyphellaria* species (p. 65), but the holes (called pseudocyphellae) in the lower surfaces of the latter have no rims. They both contain species of the blue-green alga *Nostoc*.

2 **Sticta sylvatica** grows in the same kinds of places as *S. limbata*, but the brown, branching lobes are longer and narrower and rather leathery in texture. The densely and softly hairy under-surface is darker brown and perforated with cyphellae, and short, black, rod-like outgrowths (isidia) grow in rows along ridges on the upper surface (2A).
S. fuliginosa is similar but darker brown, and the isidia are irregularly scattered on the surface. It usually consists of a single lobe, the size of a penny or a little larger, with a few smaller lobes, and it also smells rather unpleasantly of fish.

3 **Collema furfuraceum** is found on trees in districts where rainfall is high. It has rather thick, dark greenish-brown, ridged lobes, with a jelly-like texture, covered in black outgrowths (isidia) which are sometimes branched (3A). It contains a species of the blue-green alga *Nostoc*.
C. nigrescens is similar, but the isidia are rounded, and it usually has numerous small, dark-brown, cup-shaped, spore-producing structures (apothecia), which are rarely found in *C. furfuraceum*. Other species of *Collema* are described on p. 74.

4 **Lobaria pulmonaria** (Lungwort) is now restricted to western districts, but in the past it was widespread throughout Britain, though probably never very common. It forms large sheets attached at one point only to the bark of trees. It is dark green when wet and greenish-brown and papery when dry, and it is beautifully lobed and marked with regular shallow depressions. The underside (4A) is pale orange-brown with dense, short, brown hairs in the grooves. The spore-producing apothecia are red or reddish-brown discs, and powdery reproductive structures (soredia) or small outgrowths (isidia) sometimes develop along the ridges on the upper surface.

5 **Lobaria laetevirens** grows on trees in rainy districts and forms large, rather loosely attached patches of branching lobes, with neat, rounded, smaller lobes along the margins. The plant is green when wet and pale greenish-brown when dry. The spore-producing apothecia have pale red or reddish-brown discs with paler inturned margins. The general appearance of the plant is like that of a species of *Parmelia* (p. 165), but the underside is densely and softly hairy instead of being smooth and shining.

6 **Lobaria scrobiculata** is found in western Britain on trees, or sometimes on rocks or soil. It is grey with a bluish tint when wet, and pale greenish-brown when dry. The upper surface is ridged, and there are bare patches amongst the dense, short, brown hairs underneath. The dark-red spore-producing apothecia are small with thin grey margins, and clusters of pale bluish-grey, powdery, reproductive structures (soredia) develop along the margins and the ridges of the surface. The lichen contains a species of the blue-green alga *Nostoc* instead of the green alga *Trebouxia*, as other species of *Lobaria* do (*see* p. 66).

7 **Lobaria amplissima** is becoming a rare plant of western districts, though it used to be more widespread. It forms extensive patches, greenish-grey when wet and pale brownish-grey when dry, of rather thick, fairly long and broad, branching lobes with a leathery texture. The underside (7A) is covered with short, soft, brown hairs. Dark-brown, densely-branched structures containing a species of blue-green alga develop on the surface. In some parts of the world, but not apparently in Britain, these structures can grow quite separately from *L. amplissima*, and are then regarded as another lichen called *Dendriscocaulon umhausense*.

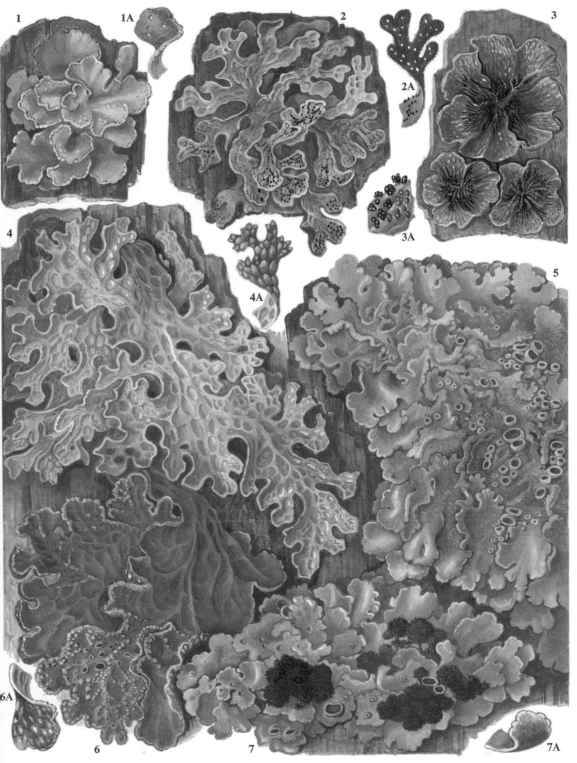

1 Sticta Limbata   2 Sticta Sylvatica   3 Collema Furfuraceum

4 Lobaria Pulmonaria   5 Lobaria Laetevirens

6 Lobaria Scrobiculata   7 Lobaria Amplissima

For a general note on lichens see page 66.

1 **Mycoblastus sanguinarius** forms light-grey, rough, warted crusts on the bark of trees, and sometimes on rocks, in upland districts, especially in Scotland. The spore-producing structures (apothecia) are black and dome-shaped (1A). In places where the outer layers (cortex) are cracked or rubbed off, the inside, blood-red tissues of the plant (medulla) can be seen.

2 **Buellia canescens** is very common on the bark of trees, often growing in shade, and it grows equally commonly on rocks and walls. It is shown on brick on p. 79. It forms smooth, very closely-attached patches with lobed margins. When dry, as shown here, it is white or greyish-white with a faint bluish tinge, and it changes to pale green on wetting. The centre of the plant is often covered with patches of white, powdery, reproductive structures (soredia).

3 **Physcia tribacia** grows on trees, rocks, and walls, especially in the south and west of Britain. It forms small, quite closely-attached patches of small whitish- or bluish-grey lobes with fan-shaped ends, which are themselves neatly lobed at the margins. Powdery reproductive structures (soredia) grown in globular clusters along the edges. Other species of *Physcia* are described on pp. 76 and 164.

4 **Xanthoria parietina** is very common everywhere on trees, walls, and roofs, but especially where the air is laden with dust that contains mineral salts. It has deep yellow or orange-yellow lobes, and the spore-producing apothecia are a deeper shade of the same colour. Another variety of this plant is shown on p. 3 and a similar species, *X. aureola*, on p. 77.

5 **Lecidea scalaris** grows in small patches on trees and fences, and very occasionally on walls. It consists of crowded, overlapping, small, thick, brownish-grey scales, which are covered with powdery reproductive structures (soredia) on their undersides. The spore-

producing apothecia (5A) are dark brown or almost black, with a dusting of a very fine powder (pruina) on the surface.

6 **Pertusaria pertusa** is common on the bark of trees throughout Britain, and forms rough, grey crusts, with a zoned margin and a white line round the edge. The spore-producing apothecia are imbedded in groups of two or three in rounded warts on the surface of the plant, and when ripe, their discs appear as small dot-like openings. A form *rupestris* is sometimes found on walls.

7 **Pertusaria hymeniea** has a rough and warted surface, shown enlarged in 7A, and its greenish-grey crusts are common on trees. The spore-producing apothecia are imbedded in the warts, and the discs, appearing as small pores at first, open out more widely later.

8 **Pertusaria amara** forms grey, granular crusts on bark, and is very common, though spore-producing apothecia are rare. The surface is covered with clusters of white, powdery, reproductive structures (soredia) which have a very bitter taste. A form *flotowiana* is sometimes found on sandstone rocks.

9 **Pertusaria albescens** has grey, coarsely-granular crusts, with a zoned margin and a white line round the edge. It grows on old trees. White, powdery, reproductive structures (soredia) grow in large, disc-shaped, raised clusters (9A).

10 **Pertusaria multipuncta** forms rather thin, grey, granular crusts on bark. The white, powdery, reproductive structures (soredia) grow on top of small warts in which the spore-producing apothecia are imbedded, usually in pairs.

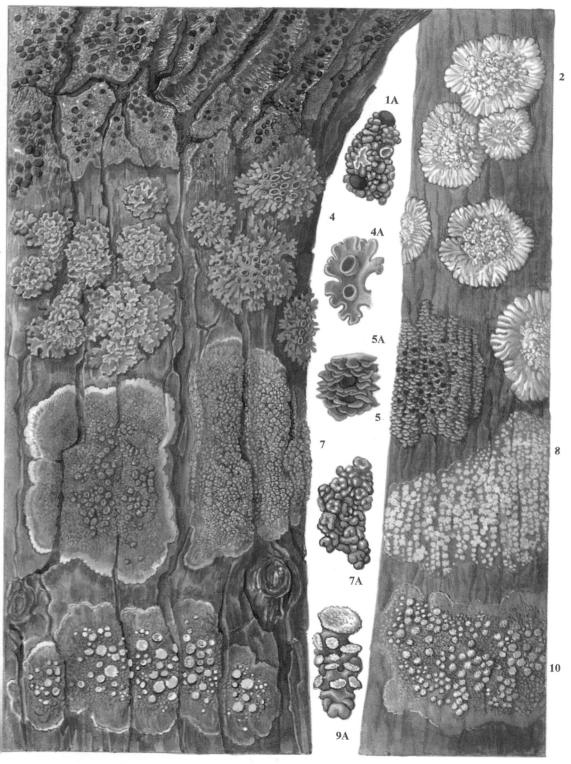

1 Mycoblastus Sanguinarius   2 Buellia Canescens   3 Physcia Tribacia
4 Xanthoria Parietina   5 Lecidea Scalaris   6 Pertusaria Pertusa
Pertusaria Hymenea   8 Pertusaria Amara   9 Pertusaria Albescens   10 Pertusaria Multipuncta

For a general note on lichens see page 66.

1   **Opegrapha atra** forms thin, pale-grey crusts on smooth bark, and is common. The very numerous, black, spore-producing apothecia (1A) have thin slit-like openings with smooth margins, and are called lirellae.

2   **Pyrenula nitida,** which grows on smooth bark, has very thin, smooth, shining, greenish-yellow crusts with black edges. When several plants are growing touching one another, the black lines are especially conspicuous and look rather like boundaries drawn on a map. Spores are produced in flask-shaped structures (perithecia) imbedded in the crust, and look like black dots on the surface. A variety *nitidella*, in which the perithecia are half the diameter of the typical form, is often found.

3   **Enterographa crassa** forms smooth, grey-green, rather thick crusts on the bark of trees. The surface is frequently intersected by black lines, giving it a map-like appearance. The minute spore-producing apothecia are sunk into the crust and appear on the surface only as tiny dots or thin, thread-like lines in rows near the margins of the lobes.

4   **Graphis elegans** is common on trees in southern parts of Britain, forming thin, pale-grey crusts on the bark. The numerous, black, spore-producing apothecia, called lirellae (4A), have slit-like openings, and the margins are deeply furrowed, so that there appear to be four 'lips' instead of two.

5   **Graphis scripta** form **serpentina** is especially common in Scotland, but is found throughout Britain, forming thin, grey crusts on bark, with numerous, black, slit-like, spore-producing apothecia, called lirellae. The margins are not furrowed as in *G. elegans*, and in this form the lirellae are variously curved.

6   **Graphis scripta** form **recta** is shown here on the same branch as form *serpentina*, though it would not, in fact, be found on the same branch, for the arrangement of the spore-producing apothecia (lirellae) in straight, parallel lines rather than in irregular, curved patterns is due to the more rapid growth of the tree during the development of the lichen.

7   **Arthonia radiata** forms thin, brownish-grey crusts, and is common on smooth bark. The numerous, small, black, spore-producing apothecia (7A) are irregularly star-shaped or lobed and have no margins.

8   **Thelotrema lepadinum** is frequent on trees in the south and west of Britain. It has a fairly thick crust with a wrinkled, shining, pale-grey surface. The spore-producing apothecia develop imbedded in round warts on the surface and, when ripe, may be seen within what look like the craters of miniature volcanoes (8A). In species of *Pertusaria* (p. 169) the apothecia are also imbedded in warts, but in groups of two or three, rather than singly.

9   **Lecidea limitata** is common everywhere on trees and fences, forming thin, grey, granular crusts, often bordered by a line. The dark greenish-brown or black spore-producing apothecia have very thin similarly-coloured margins.

10 **Leconora chlarotera** is quite common on trees, growing in small patches. It has a rather thick, wrinkled, pale-grey or greenish-grey crust, and numerous spore-producing apothecia with reddish-brown discs and crinkled grey margins. These frequently grow crowded together and then are irregularly-shaped rather than circular (10A).

1 Opegrapha Atra   2 Pyrenula Nitida   3 Enterographa Crassa
Graphis Elegans   5 Graphis Scripta *form* Serpentina   6 Graphis Scripta *form* Recta   7 Arthonia Radiata
8 Thelotrema Lepadinum   9 Lecidea Limitata   10 Lecanora Chlarotera

1 **Lycogala epidendrum** is a myxomycete (*see* p. 148), or 'slime fungus' commonly found in the summer and autumn on dead wood. The spore-producing structures are joined together in globose masses (aethelia), as shown here, and are rose red when young, but become pale yellowish brown with age.

2 **Stemonitis fusca** is another myxomycete (p. 148) frequently found on dead wood in the spore-producing stage, as shown here, in summer and autumn. The dark reddish- or purplish-brown spores develop entangled in a meshwork of fine threads growing on a short stalk (2A).

3 **Lecanora expallens** is a very common lichen on trees and fences, forming thin, whitish-yellow or yellowish-green, powdery patches. The powder consists of very small granular reproductive structures (soredia). The small pinkish-yellow, spore-producing structures (apothecia) are not often formed.

4 **Lecanora conizaeoides** forms rather thick, deeply-cracked, grey-green crusts on trees and fences and sometimes on sandstone or bricks. It flourishes especially well in towns, where it is often the only lichen growing on wood. The spore-producing apothecia (4A) have yellowish-green discs and thick, often rather irregular margins. Powdery reproductive structures (soredia) are scattered over the surface of the plant.

5 **Lepraria incana** is a very simple lichen which forms loose, soft, pale bluish-grey masses (5A) in damp, shady places. It can only grow in a moist atmosphere, but it also needs shelter from rain. It is one of the very few lichens which cannot endure direct sunlight.

6 **Pleurococcus vulgaris** is a single-celled green alga which forms a soft green powder on trees, fences, and walls wherever conditions are moist enough. Thus it is often found in rain tracks on tree trunks and on neglected buildings where drips fall from leaking gutters. It differs in colour and texture from *Lecanora conizaeoides*, which is grey green and granular rather than bright green and soft.

7 **Hormidium nitens** is a green alga which forms patches of bright-green shining threads of filaments near the base of tree trunks, or sometimes on bare, moist, but well-drained soil.

8 **Orthotrichum diaphanum** is a moss which forms small cushions, usually on trees or fences, but sometimes on wall tops or in cracks between paving stones. It grows well in towns. The dark-green leaves have long silvery-white hair-points at their tips, which, in some lights at any rate, give the plant a somewhat grey appearance. The margins are inrolled, and the nerves run almost to the tips. The enlarged drawing 8A shows, on the left, a ripe spore-capsule with its prominent ring of teeth turned back round the opening at the top, and on the right, an unripe capsule still covered by a loose, pleated cap or hood (calyptra).

9 **Orthotrichum lyellii** grows in small tufts on trees. The plant on the right in the enlarged drawing 9A shows how the leaves clasp the stem when dry, as they do in all British species of *Orthotrichum*. When wet they turn back, as shown in the plant on the left; this also shows the short, brown, rod-like reproductive structures (gemmae) which develop on the upper parts of the leaves. The leaves are long and narrow and faintly toothed at the tips, but they do not have hair-points. The nerves extend almost to the tips.

Orthotrichum affine is a common plant on trees. It is smaller than *O. lyellii*, and the leaves are shorter, with inrolled margins, and not toothed at the tips. There are fourteen other species of *Orthotrichum* found in Britain, many of which grow on rocks.

1 LYCOGALA EPIDENDRUM   2 STEMONITIS FUSCA

3 LECANORA EXPALLENS   4 LECANORA CONIZAEOIDES   5 LEPRARIA INCANA

6 PLEUROCOCCUS VULGARIS   7 HORMIDIUM NITENS

8 ORTHOTRICHUM DIAPHANUM   9 ORTHOTRICHUM LYELLII

173

The atmosphere in woodlands is damper than in the open partly because there is less wind and also because trees transpire a great deal of moisture into the air, and the humus-rich surface layers of the soil hold a considerable amount of water. Many flowerless plants flourish in these conditions, and some of them are able to grow on the tree trunks, often at a considerable distance from the ground, rather than on the soil; these are called epiphytes. The only British ferns which grow as epiphytes are species of *Polypodium* (p. 189). Many lichens, mosses, and liverworts are epiphytes. The alga *Pleurococcus* grows in rain-tracks on tree trunks, and other algae, such as *Hormidium* (p. 173), are found on the base of trees, which is considerably wetter than the higher parts, and is often covered with plants, as shown here.

1 **Dicranoweissia cirrata** grows in cushions often in drier situations than most other epiphytic mosses; it is especially common in lowland areas. The leaves are very much twisted and crisped when dry, but when moistened they untwist and bend backwards (1A). They are long, narrow, and pointed, with incurved margins and nerves that extend almost to the tips. The spore capsules are bright green when young, but ripen to pale brown; they have lids with curved points and grow vertically on pale yellowish-green stalks.

2 **Ulota crispa** forms cushions of moss on the branches of trees, especially in upland districts. The leaves are fairly broad at the base but narrow sharply into a long tapering point, and they curl and twist when dry. The nerves extend almost to the tips. The young spore capsules (2A) are covered with loose, very hairy, caps or hoods (calyptra). Later these fall off, and the ripe capsules (the right-hand drawing in 2A) are deeply furrowed and have a distinct 'waist'. Another species of *Ulota* is shown on p. 1.

3 **Isothecium myurum** forms wefts of main branches which produce several shorter branches towards the tips, giving this moss a tassel-like appearance. The leaves are oval, with toothed margins and nerves extending for about three-quarters of their length. The reddish-brown oval capsules grow erect on stalks of the same colour.
I. myosuroides is similar and grows in the same kinds of places, but is commoner in the west of Britain. The leaves have long narrow tips and strongly toothed margins, and the spore capsules are slightly curved and inclined at an angle on their stalks.

4 **Aulacomnium androgynum** has oblong leaves with pointed tips and nerves which run almost the whole of their length. Club-shaped, yellowish-green, reproductive structures (gemmae) grow in globular masses at the tips of extensions of the main stems (4A). A related moss, *A. palustre*, is shown on p. 91.

5 **Tetraphis pellucida** is a moss with oval leaves which have pointed tips; the nerve runs for about three-quarters of the leaf's length. Small, granular, yellowish-green, reproductive structures (gemmae) develop within rosettes of leaves at the tips of the stems (5A), from which they are dispersed by rain-splash.

6 **Cladonia coniocraea** is a very common lichen which sometimes grows on peaty and sandy soil as well as on trees. The slender, pointed stalks, which are hollow inside and usually curved, grow from small, grey-green 'leaves' or squamules. The upper parts of the stalks are covered with powdery reproductive structures (soredia). Other species of *Cladonia* are shown on pp. 29, 45, and 75.

7 **Lepidozia reptans** is a liverwort forming wefts of fine, branching stems. The main leaves are curved under at the tips and are divided into three or four lobes at the ends (7A). The under leaves (the right-hand shoot in 7A) are also divided in the same way. Another species, *L. setacea*, is shown on p. 89.

8 **Lophocolea cuspidata** has main leaves divided into two prominent teeth at their tips. The under leaves are very much smaller and are deeply divided. Spore capsules develop within leafy, protective sheaths (8A). This liverwort is much like *L. bidentata* (p. 39), but the leaves are symmetrical, and the teeth are longer. *L. heterophylla* is similar to *L. cuspidata* and also grows on wood, but although some of its leaves have two teeth, many of them are undivided and have rounded ends.

1 Dicranoweissia Cirrata   2 Ulota Crispa

3 Isothecium Myurum   4 Aulacomnium Androgynum   5 Tetraphis Pellucida

6 Cladonia Coniocraea   7 Lepidozia Reptans   8 Lophocolea Cuspidata

1 **Normandina pulchella** is a common lichen in the west of Britain. It consists of tiny, delicate, grey-green 'leaves' or squamules which may be distinguished from the squamules of some species of *Cladonia* (p. 174) by their narrow raised margins. Sunken, flask-shaped, spore-producing, structures (perithecia) are rare, but sometimes granular, powdery, reproductive structures (soredia) develop on the surface. This species usually grows on other small plants, and is frequently associated with *Frullania dilatata*, as shown here.

2 **Pannaria pezizoides** forms patches of bluish-grey lobes or squamules overgrowing mosses or other small plants. The spore-producing apothecia have reddish-brown discs with paler granular margins. The algal partner in this lichen is a species of the blue-green alga *Nostoc* (*see* p. 66).

Psoroma hypnorum is very like *Pannaria pezizoides*, but it contains a green alga rather than a blue-green one. It forms patches of greenish- or yellowish-brown lobes or squamules on mosses. The spore-producing apothecia have reddish-brown discs with granular margins.

3 **Frullania dilatata** is a common liverwort on the trunks of trees and sometimes on rocks throughout Britain. The leaves which clothe the regularly-branched stems each have a large round lobe and a smaller lobe in the form of a helmet-shaped 'pitcher' (3A). There are also small underleaves divided at the tips into two sharp points. The 'pitchers' contain water, and are frequently occupied by microscopic animals belonging to the group Rotifera and called *Callidina symbiotica*. It is possible that the 'pitchers' are concerned with the nutrition of the plant in the same way as may be the case in *Pleurozia purpurea* (p. 90).

4 **Frullania tamarisci** is very similar to *F. dilatata*, but it has a glossy appearance and is more often found on rocks, although it also grows on trees. It is usually a rather larger plant, and the 'pitchers' are smaller in proportion to the main lobes of the leaves. The under leaves have two rounded lobes, rather than sharp points.

5 **Hypnum cupressiforme** var. **filiforme** differs from the typical variety *cupressiforme* (No. 7) of this very variable moss in being sparingly branched and having long, fine, thread-like stems hanging downwards at fairly high levels on the trunks of trees. The shape and arrangement of the leaves (5A) are the same as in the variety *cupressiforme*, but on a more miniature scale.

6 **Hypnum cupressiforme** var. **resupinatum** forms loose, silky-textured tufts on the trunks of trees at lower levels than the variety *filiforme*. The leaves (6A) are spreading instead of being bent downwards, as in the other varieties.

7 **Hypnum cupressiforme** var. **cupressiforme** forms neat mats on the base of trees, fallen logs, rocks, walls, and soils of all kinds. It is very regularly branched, and the overlapping leaves are curved and turned downwards (7A). Each leaf is oval and tapered to a fine point, which may be lightly toothed. It has no nerve. Slightly-curved spore capsules with beaked lids grow on rather short red stalks. Other varieties of this moss are shown on pp. 43 and 49.

8 **Dimerella lutea** is a lichen which forms very thin and scanty whitish-grey crusts, usually on mosses. In the picture here it is growing on *Hypnum cupressiforme*, No. 7, and is practically invisible, except for the spore-producing apothecia. These have bright orange-yellow discs with paler margins (8A).

1 Normandina Pulchella   2 Pannaria Pezizoides   3 Frullania Dilatata   4 Frullania Tamarisci
5 Hypnum Cupressiforme *var.* Filiforme   6 Hypnum Cupressiforme *var.* Resupinatum
7 Hypnum Cupressiforme *var.* Cupressiforme   8 Dimerella Lutea

177

The majority of mosses either have fairly short upright stems and grow close together, forming extensive turfs or smaller patches called cushions, or they have creeping, branching stems which form mats. Shady, moist places in woodlands, however, provide a habitat for two rather more luxuriant types of growth form. The first four mosses on this page are 'tree-like' (dendroid), with creeping stems along the surface of the ground from which grow short upright stems, ending in a tuft of branches. The rest on this page form wefts, with stems that neither creep closely along the surface of the ground nor grow upright, but form deep, loose tangles over considerable areas.

1 **Mnium undulatum** forms extensive light-green patches in moist, shady places, sometimes growing amongst grass. The curved branches at the tips of the main stems give the plants a 'palm tree' appearance. Creeping runner-like shoots, and upright, curved, unbranched stems are also formed. The long, wavy leaves are oblong and have toothed margins and nerves which run almost to the tips (1A). Other species of *Mnium* are described on p. 180.

2 **Thamnium alopecurum** has dark, reddish-brown to almost black stems, and dark green leaves. Those on the upright stems are broad and triangular, while those on the branches are oblong. Both types of leaf are pointed and toothed at the tips. This species is found both in very wet shady places and in drier situations in woodlands on chalky soils. Sometimes the plants become detached and roll about on the ground, behaving rather like the flowering plant 'Tumbleweed' of western North America. In this respect it also resembles *Leucobryum glaucum* (p. 181).

3 **Climacium dendroides** has stout, upright stems and small oval leaves with hooded tips. The leaves on the branches are longer and more crowded and have pointed tips and toothed margins. This moss grows in woodlands and beside lakes, often amongst grass.

4 **Rhodobryum roseum** usually grows amongst grasses and other plants, and has rosettes of broad, delicate leaves at the tops of upright stems. The margins

are toothed towards the pointed tips of the leaves, and the nerves run almost the whole length.

5 **Rhytidiadelphus loreus** has red, branching stems, clothed with broad leaves which are narrowed abruptly into a long, turned-back point (5A). There are short, double nerves, and the margins of the leaves are lightly toothed. Other species of *Rhytidiadelphus* are shown on p. 43.

6 **Ptilium crista-castrensis** grows especially in conifer woods in Scotland, and has very regularly-branched stems clothed in golden-green leaves with long, curved tips and small, branched, hair-like outgrowths (paraphyllia).

7 **Thuidium tamariscum** has slender, dark-green or almost black stems, clothed in broad, triangular leaves with long points, and small, branched, hair-like structures (paraphyllia). The main stems are branched three times in a very regular manner, and the leaves on the branches are smaller and narrower than those on the main stems (7A).

8 **Hylocomium splendens** has red stems clothed in oval leaves with long points and toothed margins. Amongst the leaves are numerous small scale-like outgrowths (paraphyllia). The main stems are branched in a very regular manner, and have a glossy, golden-green 'ostrich plume' appearance.

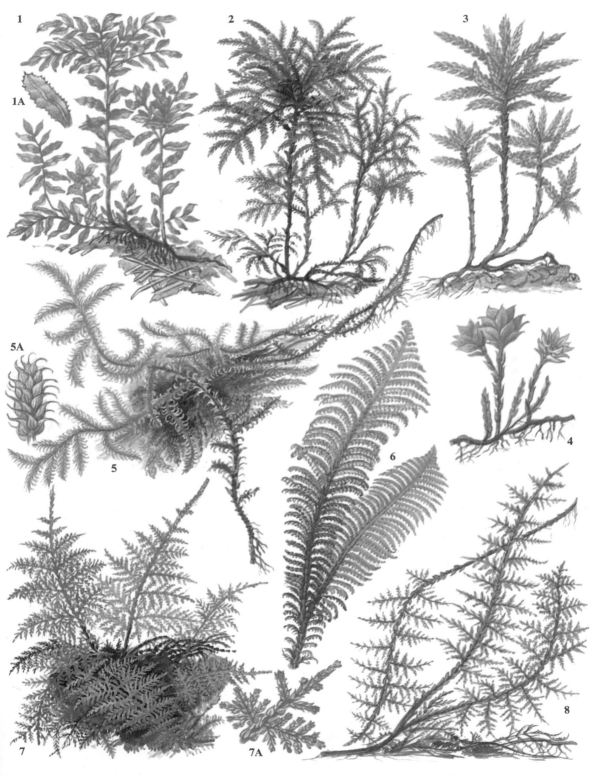

1 Mnium Undulatum   2 Thamnium Alopecurum   3 Climacium Dendroides
4 Rhodobryum Roseum   5 Rhytidiadelphus Loreus   6 Ptilium Crista-castrensis
7 Thuidium Tamariscum   8 Hylocomium Splendens

1   **Mnium hornum** is one of the commonest of all mosses in woodlands, and frequently grows on wood and peat, forming extensive, rather dull dark-green turfs. In spring the very light-green young shoots make a conspicuous contrast. The oblong and pointed leaves have nerves extending almost to the tips and pairs of teeth along the margins (1A). The spore capsules grow on long stalks curved like the neck of a swan. Male plants have a rosette of leaves with a dark centre at the tips of the stems.
M. affine is similar but has almost round leaves, with the nerve projecting in a point, and single teeth along the margins.
M. longirostrum is like *M. affine* except that the points and teeth of the leaves are shorter, and the capsule has a long beak-shaped lid. *M. undulatum* is shown on p. 179.

2   **Mnium punctatum** grows in wet, shaded places in woodlands, by streams, and on mountains. The large oval pale-green leaves develop a reddish tinge with age; they have well-developed nerves and prominent borders, and when held to the light have a dotted appearance (2A). The spore capsules grow on orange stalks and hang downwards. Male plants have open rosettes of leaves at the tips of the stalks.
M. pseudopunctatum is similar, but there is no border to the leaf.

3   **Leucobryum glaucum** forms very large cushions in woodlands and on wet moorlands. The leaves are grey green, unlike those of any other moss, and become whitish green when dry. Some of the cells in the leaves are hollow and have small pores in the walls through which they absorb water; in this respect they are like *Sphagnum* (p. 86), although they differ very much in general appearance. The cushions are very loosely attached and are easily rolled about and broken up; the only other British moss which continues to grow when detached in this manner is *Thamnium alopecurum* (p. 179). Unfortunately, its attractive appearance and the ease with which it is gathered causes it to be collected for sale as a decoration, and it is seriously threatened with extinction in many areas.

4   **Atrichum undulatum** ('Catherine's Moss') is common in woodlands on all but the poorest or most chalky soils. The long, narrow, rather dark-green leaves have wavy margins, and there are from 4 to 7 thin vertical plates running along the prominent nerves. The tip of each leaf is pointed, and there are pairs of teeth along the edges (4A). The curved spore capsules are held at an angle on long stalks and have lids with very long beaks. The 18th-century botanist Friedrich Ehrhart associated this moss with Catherine the Great of Russia.

5   **Fissidens adianthoides** grows chiefly in wet places in upland areas. The leaves are in two ranks, and each has a clasping 'pocket' at the base (5A). The nerves do not run quite to the tips, which are slightly toothed, and there is a faint, pale band along the margins. The spore capsules are held almost upright on long, red stalks.

6   **Fissidens bryoides** is common in moist, shady places and, like all species of *Fissidens*, it has leaves with basal 'pockets' in two ranks. The well-developed nerves extend right to the tips of the leaves, which have prominent, thickened borders. The stalks of the spore capsules grow from the tips of the shoots (6A).

7   **Fissidens taxifolius** has rather wider leaves than *F. bryoides*, but they do not have thickened borders. The nerves extend from the tips as short points. The stalks of the spore capsules grow from near the base of the plant (7A). There are about twenty species of *Fissidens* in Britain, but many of them are rare or very small and easily overlooked.

1 Mnium Hornum   2 Mnium Punctatum
3 Leucobryum Glaucum   4 Atrichum Undulatum
5 Fissidens Adianthoides   6 Fissidens Bryoides   7 Fissidens Taxifolius

1 **Eurhynchium praelongum** is a common moss growing at the base of trees in woodlands, on clayey banks, and in hedgerows, and is very tolerant of deep shade. The long, creeping main stems are clothed in broad, heart-shaped leaves, which narrow at the base and contract suddenly into narrow points at the tips. The main stems are rather regularly branched, and the leaves on the branches are narrower and much smaller than those on the main stems. All the leaves have toothed margins and nerves that run as far as the middle, or a little further. This species is very similar to *E. swartzii* (p. 45), but the latter is less regularly branched, and the leaves on the branches are nearly as large as those on the main stems.

Eurhynchium confertum grows on the ground in woodlands, and on stones, rocks, and walls in shady places. It forms small patches of irregularly-branched stems, clothed in pointed, oval leaves with minutely-toothed margins and nerves which run to just beyond the middle. It is very similar to *Brachythecium velutinum* (p. 39), which, however, has longer and narrower leaves with longer, finer points. The spore capsules are curved, as in other species of *Eurhynchium*, and have distinctly beaked lids with smooth, pale-brown stalks.

2 **Cirriphyllum piliferum** forms loose mats or wefts of moss, with regularly-branched stems in woodlands and grassy places. The oval leaves narrow abruptly at the tips to form pale-green hairs (2A). The nerves run to just beyond the middle, and the leaf margins are very lightly toothed.

3 **Dicranum scoparium** forms mats or cushions of moss in woods and on heaths. The leaves are long and narrow, tapering into long, jaggedly-toothed points into which the nerves run, and they are usually turned to one side, though sometimes not conspicuously so. The spore capsules are curved and have lids with long beaks.
D. majus has leaves about twice as long as *D. scoparium*, with sharp, finely-toothed points; they are very much curved to one side. The curved spore capsules have pale yellow stalks and grow in clusters of up to six.

4 **Dicranella heteromalla** is a common moss everywhere, except on chalky soils, and forms soft, bright-green cushions. The narrow leaves taper into long, fine points, and are curved to one side. They have broad nerves running their whole length. The slightly curved, pear-shaped spore capsules are held almost horizontally on yellow stalks. A related species, *D. squarrosa*, is shown on p. 59.

5 **Isopterygium elegans** is a moss of sandy, but not chalky, soils, forming mats of somewhat flattened, rather sparingly-branched shoots. The leaves are oval or oblong with finely-pointed tips, and have either very short nerves or none at all. Bunches of thread-like shoots with widely-spaced, very tiny leaves often grow amongst the leaves on the main stems; these become detached and are able to grow into new plants.

6 **Plagiothecium denticulatum** is a common moss of woodlands. It has very much flattened shoots, and the leaves, which are rather widely spaced, are oval with sharp, pointed tips and with double nerves extending not quite as far as the middle. The spore capsules grow almost upright on orange-red stalks.
P. undulatum is a larger plant with very much flattened, whitish-green shoots. The leaves are like those of *P. denticulatum*, but are larger and markedly wavy or wrinkled.

7 **Scapania nemorosa** is a liverwort which grows in shady places in woodlands. Each leaf has two lobes, one more than twice as large as the other (7A), with fine, hair-like teeth on the margins. Clusters of granular reproductive structures (gemmae) sometimes grow at the tips of the shoots. A related species, *S. undulata*, is shown on p. 59.

8 **Plagiochila asplenioides** is a common liverwort of moist, shady places. It has large oval leaves which sometimes have toothed margins. The spore capsules develop within protective sheaths, and when they are ripe, are carried upwards on pale greenish-brown stalks. A variety *major* is quite common, which is much larger than the ordinary form, and has overlapping leaves.

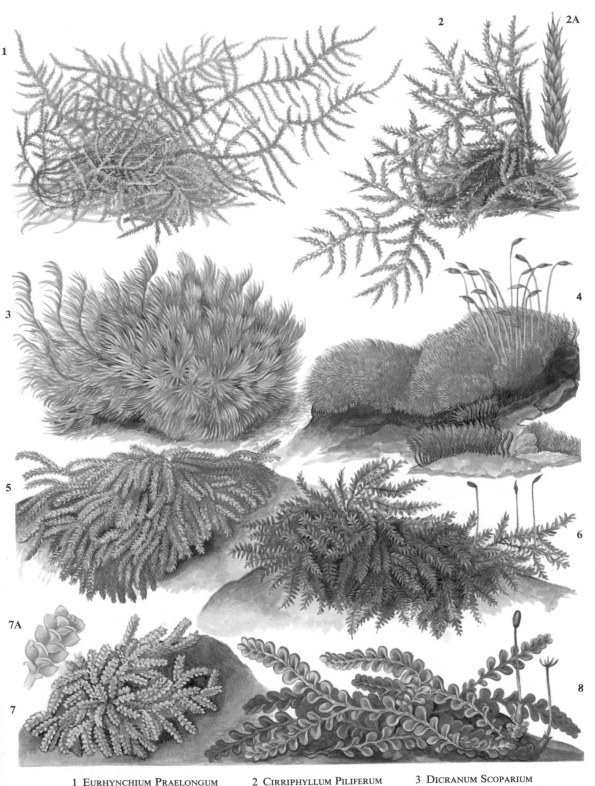

1 Eurhynchium Praelongum   2 Cirriphyllum Piliferum   3 Dicranum Scoparium
4 Dicranella Heteromalla   5 Isopterygium Elegans   6 Plagiothecium Denticulatum
7 Scapania Nemorosa   8 Plagiochila Asplenioides

1 **Pteridium aquilinum** (Bracken) is the most widespread of all ferns and is found in every continent except Antarctica, growing abundantly in the tropics as well as in temperate regions right up to the arctic circle. In Britain it is common on heaths, especially on light, acid soils, and in woodlands, where the fronds may grow taller than a man. It is a serious pest in grassland and is difficult to eradicate. The plant has an extensive system of branching underground stems, with long, thick shoots growing deep in the soil, and thinner, shorter branches running near the surface. The fronds grow singly in two rows at intervals from the thinner underground stems, and have tall, tough, vertical, channelled stalks with large, much-divided, triangular blades folded back at the tops of them. The very young fronds are clothed with chaffy brown scales and a dense coat of soft hairs, which soon fall off. The illustration shows the blade of a slightly older frond with some of the lobes unrolled and others still tightly coiled. Soon after this stage, glands at the base of the main branches of the frond, called 'extra-floral nectaries', secrete a sugary liquid which is much sought after by ants and other insects. The detail drawing of the upper surface of part of a frond (1A) shows the bluntly-tipped oblong lobes and the channelled stalks. Spore cases develop in a continuous row all along the underside of the margins of the lobes (1B), and are protected by the inrolled edge of the leaf, which is fringed with hairs.

2 **Dryopteris aemula** ('Hay-scented Fern') grows in shady, warm and moist situations, especially in deeply-wooded valleys, or sheltered by other more hardy ferns. It is common in west Scotland, western Ireland, and south-west England, rare in Wales and the Lake District, and locally plentiful in the weald of Kent and Sussex. Outside Britain, it is found only in western France and in the Azores. The fronds, which remain green through the winter, have minute glandular hairs on both surfaces which contain an aromatic substance called coumarin and give the

plant a pleasant smell when crushed. They have brownish-green stalks clothed in pale-brown, narrow, pointed scales, and blades with from 15 to 20 pairs of main lobes, the lowest pair being the largest. These are divided into from 10 to 15 pairs of smaller lobes, which are further divided. The lobes are toothed at the margins and have short, bristle-like hairs at the tips (2A). The bottom lobe on the lower side of the lowest main lobe is markedly larger than the others. The margins of the lobes are upturned, giving the frond an attractive crisped appearance. Spore cases are produced in groups, each protected by a kidney-shaped scale (indusium), on the undersides of the fronds.

3 **Dryopteris filix-mas** (Male Fern) is very common in woodlands in all parts of Britain. The stalks of the fronds are clothed with pale-brown scales, and the blades have from 20 to 35 main lobes, the longest ones being near the middle. Each main lobe is deeply divided into smaller lobes, which are toothed at the margins all the way round (3A). The spore cases develop in clusters on the undersides of the fronds, each protected by a kidney-shaped scale (indusium) which becomes grey when the spores are ripe (3B).

4 **Dryopteris borreri** is widespread in Britain, but is commonest in the north and west and in the weald of Kent and Sussex. It is similar to *D. filix-mas*, but the stalk is more thickly covered with orange-yellow to golden-brown scales, and there is a dark spot at the base of each main lobe on the underside where it joins the stalk. The smaller lobes have the appearance of being 'cut off' at the tips, and are not toothed at the sides. The groups of spore cases shown in 4A are young, and the protective scales have not yet turned grey. Other species of *Dryopteris* are described on pp. 94 and 186.

1 Pteridium Aquilinum

2 Dryopteris Aemula   3 Dryopteris Filix-mas   4 Dryopteris Borreri

FRONDS – HALF LIFE SIZE   DETAILS – LIFE SIZE

In woodlands of all kinds the soil is profoundly affected by the presence of the trees. The accumulation of dead leaves and twigs on the ground increases the amount of humus in the soil, although the rate at which different kinds of leaves decay varies widely. Water movements in the soil are also affected, for trees absorb very large amounts of water and transpire it to the atmosphere; also the leaf canopy and the leaf litter on the ground intercept rainfall and alter the pattern of run-off. In turn, the changed pattern of water movement alters the distribution of minerals in the soil. Woodland soils are usually moist, rich in minerals, and tend to be acid, and these conditions are favourable to the growth of most ferns.

1   **Dryopteris dilatata** ('Buckler Fern') is common in woodlands throughout Britain. The fronds usually form a crown-like tuft, and their stalks are clothed in pale-brown pointed scales which have a dark-brown or almost black central stripe (1A). The blades have up to 25 pairs of main lobes, the lowest pair forming a roughly triangular outline. The main lobes are divided into from 10 to 20 smaller lobes, which are further divided, and are toothed at the margins, with short, stiff, bristle-like hairs at their tips (1B). The groups of spore cases on the undersides of the fronds (1C) are protected by kidney-shaped scales (indusia).

Dryopteris villarii is rather like *D. dilatata:* it has clusters of spore cases covered with kidney-shaped scales (indusia), but the teeth of the lobes do not have bristle-like hairs at their tips. The fronds are covered with small, yellow, glandular hairs which give them a pleasant balsam-like fragrance when crushed. It is a rare plant, found in crevices in limestone rocks in north Wales and northern England.

2   **Dryopteris carthusiana** ('Prickly Buckler Fern') is widespread in damp places on heaths and in woodlands. The rather delicate fronds grow in small tufts from short, creeping rootstocks. The slender stalks are sparsely covered with pale-brown, pointed, oval scales and are about the same length as the blades. These have from 15 to 25 pairs of main lobes, the lowest three or four pairs being about equal in length, and longer than the rest. They are further divided into from 7 to 12 pairs of smaller lobes, with

toothed margins and short, stiff, bristle-like hairs at their tips. The bottom leaflet on the lower side of each of the lowest pair of main lobes is markedly larger than any of the others. Spore cases develop on the undersides of the fronds in groups, each protected by a kidney shaped scale (indusium) (2A). Other species of *Dryopteris* are described on pp. 94 and 184.

3   **Polystichum setiferum** ('Shield Fern') grows in woods and on hedgebanks in the south and west of Britain. The fronds grow in basket-like crowns from a stout rootstock. The stalks are covered with brown, oval, pointed scales, and the blades have up to 40 pairs of main lobes, divided into smaller lobes with toothed margins and long hairs at the tips of the teeth. The margins at the base of the smallest lobes meet at right angles. Spore cases are produced on the undersides of the fronds in numerous separate clusters, which are protected by circular scales (indusia) attached by a central stalk (3A).

4   **Polystichum aculeatum** ('Prickly Shield Fern') is found in woods and hedgerows throughout Britain, though it is rather uncommon in Ireland. The fronds are stiffer and more leathery than those of *P. setiferum*. They usually have more pairs of main lobes, and the hairs on the teeth at the margins are short and bristle-like (4A). The margins at the base of the smallest lobes meet in angles that are considerably less than right angles. The spore-case clusters are protected by circular scales (indusia) with central stalks. Another species of *Polystichum* is described on p. 54.

1   1   1C   1B   2   2A   3A   1A   3   4   4A

1 Dʀyopteʀis Dilatata   2 Dʀyopteʀis Caʀthusiana
3 Polystichum Setifeʀum   4 Polystichum Aculeatum
FRONDS – QUARTER LIFE SIZE   *DETAILS – LIFE SIZE*

1   **Polypodium interjectum** (Polypody) grows amongst mosses on trees, in crevices in rocks, and on walls and hedgebanks throughout most of Britain. It has a sparingly-branched, somewhat flattened, creeping stem, clothed with small, reddish-brown, pointed scales. The smooth-stalked fronds, which remain green through the winter, grow singly at intervals in two rows along the stem. They are oval in outline and divided into from 5 to 25 pairs of lobes, with clusters of spore cases on the undersides (1A). The spore cases are slightly oval rather than perfectly round, and are not protected by a scale. The lowest pair of lobes projects slightly forwards, and the longest pair is usually the fifth from the base.

P. vulgare has quite flat, oblong fronds, with the lobes equal in length except at the tips, and the groups of spore cases are perfectly round.

P. australe, which is found in the south and west of Britain, has triangular fronds, the second pair of lobes being the longest, and the lowest pair projecting a little forwards. The groups of spore cases are slightly oval, and the spores do not ripen until late autumn or the following spring. Usually the margins of the lobes are lightly toothed. A variety *cambricum* ('Welsh Polypody') is sometimes found in which the lobes are deeply and elegantly divided, giving the fronds an attractive 'ostrich feather' appearance.

2   **Thelypteris limbosperma** ('Mountain Fern') is found in woods and on steep banks by streams in upland areas, throughout most of England, Wales, and Scotland, except on chalk and limestone soils; it is rather rare in Ireland. The fronds grow in tufts from a short, stout rootstock. The stalks are about a quarter as long as the blades, and are sparsely covered with small, bright-yellow, glandular hairs, which give the plant a pleasant lemon scent when crushed. Spore cases develop on the undersides of the fronds in groups, partially protected by small, irregular scales (indusia), which soon wither (2A).

Thelypteris palustris ('Marsh Fern') grows in wet places, mainly in south and east England, and has fronds rather like those of *T. limbosperma*, but they grow singly from long, creeping, underground stems, and are not covered with glandular hairs.

Thelypteris phegopteris (Beech Fern) is common in damp woods and amongst shaded rocks in Scotland, northern England, and Wales, but is rare elsewhere in Britain. The blades of the fronds are triangular in outline, and are bent backwards at the tops of slender stalks, which are nearly twice as long as the blades. The whole plant is covered with scattered, short, white hairs. The fern has no particular association with beech trees.

3   **Blechnum spicant** (Hard Fern) is found in woods and on heaths and moors throughout Britain, except on chalk and limestone soils. The fronds grow in tufts from a short rootstock. The outer spreading fronds (3A) have from 30 to 60 pairs of short lobes, set closely together, like the teeth of a comb, on either side of the main stalk, and remain green in the winter. The inner ones (3) grow erect, and have much narrower, more distantly-spaced lobes, which produce on their undersurfaces pairs of oblong groups of spore cases (3B); these fronds die away as soon as the spores are shed.

4   **Athyrium filix-femina** (Lady Fern) is common in damp woods and amongst shaded rocks throughout Britain. The tall fronds usually droop at the tips and form a graceful crown at the top of a short, stout rootstock. They have channelled stalks about a quarter the length of the blades and are clothed with pale-brown scales. There are from 25 to 30 pairs of main lobes, the longest ones near the middle, and these are divided into smaller lobes with toothed margins. There are two common forms of the plant, one with a yellowish-green midrib to the blade, as shown here, and the other with a purplish-red midrib. Spore cases are produced on the undersides of the fronds in clusters, protected by crescent-shaped scales with finely-toothed edges (4A).

1 Polypodium Interjectum   2 Thelypteris Limbosperma
3 Blechnum Spicant   4 Athyrium Filix-femina

FRONDS – HALF LIFE SIZE   *DETAILS – TWICE LIFE SIZE*

The twenty species of *Equisetum*, the Horsetails, found on the earth today are the only survivors of an ancient group of plants, the SPHENOPSIDA, which included many of the trees that grew in the coal forests 250 million years ago. All living species, such as those shown here and on p. 95, are much smaller, and have jointed stems with whorls of small, dark, tooth-like leaves united into sheaths at the joints. Although sometimes referred to as 'Fern Allies', they are not closely related to ferns. The spores, unlike those of ferns, are soft, green, and short-lived, and are produced in cones growing at the tips of the shoots.

1 **Equisetum arvense** (Horsetail) is a common plant in woods and fields, especially on light soils, and is sometimes a serious weed of cultivated land. It has a slender, tough, dark-brown or black underground stem, which produces tufts of wiry, branching roots, and numerous oval tubers, each about the size of a haricot bean. All the underground portions of the plant are densely covered with fine rusty-brown hairs. In the spring, short, pinkish-brown stems appear (1A), each with a blunt-tipped, oval cone carrying the spores; they are hollow, soft, and fleshy, unbranched, and devoid of any green colouring, and they die away as soon as the spores are shed. The much larger, green shoots appear later, and have numerous whorls of branches. The sheaths on the main stems are fairly long, though shorter than the first joints of the branch stems, and they have from 8 to 20 short, black-tipped, pointed teeth (1B). There are an equal number of grooves on the stems corresponding to the teeth of the sheaths, and the ridges between are moderately rough to the touch. The central hollow cavity is less than half the diameter of the stem, and there is a ring of smaller cavities alternating with the grooves.

Equisetum sylvaticum is a handsome plant of woodlands, especially in the north of Britain. It has whorls of drooping branches, which are themselves further branched, giving the plant an attractive feathery appearance. The sheaths on the main stems are long and green, with the teeth united to form from 3 to 6 brown, broad, pointed, oval lobes, each with several ribs.

2 **Equisetum telmateia** is widespread in Britain, except in north-east Scotland. The dark-brown underground stems are long and stout, and the younger parts are covered with very fine, short, brown hairs. Tufts of rather thick, much-branched roots and pear-shaped tubers, each about the size of a pigeon's egg, are produced. Cone-bearing stems (2A) appear in the

spring, with prominent sheaths which are white at the base and brown above, and have 20 to 30 long, dark-brown, almost hair-like teeth. The ordinary stems, which are shown here in both the immature and fully-developed stages, appear later. They are white with green sheaths and whorls of long, drooping branches. The sheaths (2B) are considerably longer than the first joint of the branches, and have 20 to 30 long, brown, finely-pointed teeth. The hollow cavity in the stem occupies about two-thirds of the diameter.

3 **Ophioglossum vulgatum** (Adder's Tongue) is a fern which grows amongst open scrub and in grassland throughout Britain, but is less common in the north and west. Each plant has a short, upright rootstock, with several fleshy, usually unbranched, horizontally-growing roots, which sometimes give rise to new plants at their tips. A single frond grows each year, and this divides to form a pointed, oval leaf and a straight, unbranched stalk carrying up to forty pairs of spore cases, and ending in a sharply-pointed tip. A related plant, *Botrychium lunaria*, is described on p. 56.

4 **Phyllitis scolopendrium** (Hart's Tongue) is the only British fern in which the fronds are normally quite undivided. It is very common throughout Britain, except in north-east Scotland. The fronds are long and strap-shaped, tapering to the tip, and heart-shaped at the base. They have a prominent midrib, and the stalks are clothed with narrow, pointed, brown scales. Spore cases are produced in paired oblong groups along the branch veins on the undersides of the fronds, each group being protected by a long, narrow scale (indusium). Plants with fronds which are forked at the tip or otherwise lobed or divided are not uncommon and are prized by gardeners who specialize in ferns, although they are not as popular in cultivation as they were 100 years ago.

1 Equisetum Arvense   2 Equisetum Telmateia
3 Ophioglossum Vulgatum   4 Phyllitis Scolopendrium
FRONDS – QUARTER LIFE SIZE   DETAILS – LIFE SIZE

CLASSIFICATION OF THE MAIN GROUPS OF FLOWERLESS PLANTS

For convenience, the **Classes** of plants described in this book have been arranged here in five main groups. When a large number of species in a Class has been described, the **Orders** to which they belong are listed.

FERNS AND 'FERN ALLIES'

Filicopsida (Ferns)

Sphenopsida (Horsetails)

Lycopsida (Clubmosses)

    LYCOPODIALES

    SELAGINELLALES

MOSSES AND LIVERWORTS

Bryophyta

    MUSCI (Mosses)

    HEPATICAE (Liverworts)

FUNGI

Basidiomycetes

    APHYLLOPHORALES

    AGARICALES

    GASTEROMYCETALES

    TREMELLALES

Ascomycetes

    PEZIZALES

    TUBERALES

    HELOTIALES

    SPHAERIALES

ALGAE

Chlorophyceae (Green Algae)

Phaeophyceae (Brown Algae)

    ECTOCARPALES

    FUCALES

    LAMINARIALES

Rhodophyceae (Red Algae)

    GIGARTINALES

    CRYPTONEMALES

    RHODYMENIALES

    CERAMIALES

Cyanophyceae (Blue-green Algae)

OTHER GROUPS

Lichens
Myxomycetes
Charophyta

THE CLASSIFICATION OF FLOWERLESS PLANTS

The first plants with flowers began to grow on the earth about one hundred million years ago at the time when the higher insects were beginning to appear. Since then, they have spread and increased until they are the dominant plants in almost every kind of habitat. Their success is partly due to the efficiency of their reproductive system, in which pollination is brought about in most cases with the unconscious assistance of bees, butterflies, and moths. The insects in return receive food in the form of nectar and small amounts of pollen. This represents a much smaller expenditure of material by the plant than would be involved in inevitable wastage if the process of pollination were left to chance. The reproduction of flowering plants culminates in the production of a most efficient type of dispersal structure — seeds. Many plants without flowers also produce seeds; some such groups existed 250 million years ago, and others, such as conifers and cycads ('sago-palms'), still flourish today, which is an indication of the success of seeds as reproductive units.

None of the plants described in this book produce seeds; they rely on spores for reproduction and dispersal. A seed is a complex structure. Even the smallest seeds consist of many cells, and the larger ones already contain a highly developed embryo before they leave the parent plant. They all contain a food store, which may be quite substantial, and have an elaborately structured seed coat which protects the contents from excessive heat and light and prevents water loss. Spores, in contrast, are comparatively simple. Each is a single cell, with very limited food reserves within it, and a cell wall which is flimsy compared with the coat of a seed. Their chances of surviving and growing into new plants are correspondingly very small, and to compensate for this they are produced in large, in some cases vast, numbers. It is advantageous to the species for the spores to be scattered over a fairly considerable area to increase the chances of some of them falling into suitable situations for germination, and to ensure that the young individuals do not have to compete with one another or with the parent plant. Equally important to this dispersal in space is dispersal in time. If all the spores are shed at once, and germinate at once, then a comparatively short period of drought, or a sudden rainstorm, or some other small-scale local disaster can easily destroy them all. The methods by which spore dispersal in both space and time is achieved in the many different kinds of spore-producing flowerless plants are diverse and in some cases complicated. They have been described for individual species or groups wherever possible in this book.

Spore plants have a far more ancient lineage than flowering plants, and some were flourishing in the earliest times we have knowledge of, when the fossil record begins about 500 million years ago. Most of them have such a delicate structure, however, that they have left only vague and scanty traces in the rocks, and we have little detailed knowledge of their history in the remote past.

FERNS AND 'FERN ALLIES'

The higher spore plants, ferns, clubmosses, and horsetails, have as elaborate a structure as the flowering plants. They have easily distinguishable stems, roots, and leaves, each with a distinct and complex internal anatomy. On the surface is a layer of cells called the epidermis, with a thin, waxy skin, the cuticle, on the outside. This is perforated at intervals by minute pores or stomata, the size of which is controlled by two protecting guard cells. The stomata lead into a system of spaces between the internal cells of the plant, which allow for an exchange of gases with the atmosphere. Inside are strands of woody tissue (xylem) which carry water from the roots to the stems and leaves. Running parallel with them are softer and more delicate tubes (phloem) which distribute sugar and other dissolved food substances. Although they all have in common this complexity of structure, they vary greatly in other ways, especially in the details of the reproductive processes, and the three main groups are not at all closely related to one another.

Filicopsida (Ferns)

Ferns have two distinct and alternating stages in their life cycle, the prothallus (*see* p. 97), which is comparatively short-lived, and the fern plant itself. The leaves of ferns are usually more like branch systems than like the leaves of other plants, and are usually referred to as fronds. When young, the various lobes of the fronds are spirally coiled tightly upon themselves, and the frond itself is also spirally coiled; it has often been likened to a bishop's crozier. Spores are produced in stalked spore cases which grow in clusters (sori) on the undersides of the fronds. In many species each sorus is covered by a protective scale (indusium). In most ferns the individual spore cases in a sorus ripen at different times, thus ensuring that the spores are shed over a period. The comparatively few British species are representatives of a number of different groups of ferns which are much more abundant elsewhere in the world. Of the twenty genera mentioned in this book, *Asplenium* and *Ceterach* (p. 73) and *Phyllitis* (p. 191) belong to a large tropical family ASPLENIACEAE, and *Dryopteris* (pp. 185,187), *Polystichum* (pp. 55, 187) and *Gymnocarpium* (p. 55) are members of another large family ASPIDIACEAE, which is found most abundantly in Asia. The remainder are distributed amongst twelve other families.

Lycopsida (Clubmosses)

Clubmosses are not very closely related to ferns, and the details of their structure and life cycle are very different. The prothallus stages are quite unlike those of ferns. The spore cases grow at the base of the upper surfaces of small, undivided leaves, rather than underneath branching fronds. There are two main groups.

LYCOPODIALES. The 100 or so species of this group are best regarded as belonging to a single genus, *Lycopodium*, although they are sometimes grouped into four different genera. Five species are found in Britain (pp. 57, 93). The prothalli grow entirely underground, and obtain their food from soil-inhabiting fungi which have entered into close associations with them. Only one kind of minute, dust-like spore is produced, and the leaves, unlike those of *Selaginella*, do not have ligules.

SELAGINELLALES. There are about 600 species of the genus *Selaginella*, of which only one (p. 57) is British. The prothalli are very tiny structures compared with those of ferns. There are two kinds of spore, the minute, dust-like microspores, and the much larger megaspores. A small scale called a ligule grows on the upper surface of the base of each leaf. *Isoetes* (p. 97) looks very different and is sometimes placed in a group of its own; but it resembles *Selaginella* in having ligules on the leaves and two kinds of spore. Several other genera, which included many large trees, are known to have existed in the remote past.

Sphenopsida (Horsetails)

Horsetails cannot be considered as at all closely related to ferns. The spores are produced in cones which have some hundreds of hexagonal scales with centrally-placed stalks, arranged in a dozen or more whorls, and each with from five to ten thin-walled spore cases on their undersides. Each spore is soft and green and has a pair of bands, called elaters, attached to it by their centres and tightly coiled spirally when moist; when they become dry they open out and aid in spore dispersal by their movements. The prothallus is green and repeatedly branched and lobed, rather than heart-shaped as in ferns. There is only one still existing genus, *Equisetum*, with about 25 species, of which 10 are British (pp. 95, 191). Several other genera, which included tree-like forms, existed some hundreds of millions of years ago.

MOSSES AND LIVERWORTS

Bryophyta

Most of the species in this group are quite small and grow in moist situations. In the Bryophyta the plant corresponds to the prothallus of ferns, and the spore-producing stage in the life cycle is merely a stalked capsule which is unable to live separately. In most cases the plants have stems and leaves, but of a simpler structure than those of higher plants, for there is no epidermis, nor air space system between the cells, nor woody tissue. They have no true roots but are anchored to the soil by special hairs called rhizoids, which correspond to the root hairs of higher plants. They are classified into two main groups.

MUSCI. The leaves of mosses are spirally arranged, or in a few cases in two rows. There are three main groups. *Sphagnum* (pp. 87, 89) has a unique 'spongy' structure and explosively-opening spore capsules. In *Andreaea* (p. 61) the spore capsules open by four slits, but the walls do not split into flaps or 'valves'. There are about 160 other genera in Britain, known collectively as *Bryales*, in which the spore capsules have quite an elaborate structure, with a little lid which falls off when the spores are ripe. The mouth of the capsule is fringed with a ring of teeth (peristome), which move in and out with changes in the humidity of the air, thus releasing the spores gradually over a period of time.

HEPATICAE. The majority of liverworts have three rows of leaves, although in a few the plant is a flat, lobed structure. The walls of the spore capsules split open into four flaps or 'valves'. There are about 80 genera in Britain (*see* p. 100).

FUNGI

Fungi differ from all other plants in that they lack the green pigment chlorophyll and are thus unable to make use of light energy from the sun to manufacture sugar and other food substances from simple materials. A few microscopic forms live on animals, but the majority obtain their food from plant and animal remains in the soil or from living plants. The actively growing and feeding part of the fungus consists of a branching weft of fine threads called hyphae. All the species described in this book develop large spore-producing structures which are built up from hyphae and not from cells, as are the parts of other plants. These larger fungi are classified into two main groups, each of which is further sub-divided.

Basidiomycetes

These are characterised by spores budded off, usually in groups of four, from special cell-like structures called basidia. When ripe, the spores are shot off violently, but in contrast to those of the other main group, the Ascomycetes, only for a very short distance (about a tenth of a millimetre or so). Consequently, although the basidia are often produced in layers lining the insides of tubes or on the surfaces of flat plates called gills placed quite closely together, the tubes or gills are always arranged vertically so that the spores, once they are shot clear of the basidial layer, can fall freely. The Basidiomycetes described in this book are classified into four groups.

APHYLLOPHORALES. In this group the spore-producing structures are of various shapes. In many species they are bracket- or shelf-like (p. 111, 117, 129), while in some they are thin, flexible, and fan-like (*Thelophora*, p. 47). In others the spore-producing layer is exposed on the surface of soft 'teeth' or 'spines' (*Hydnum*, p. 153), or on club-shaped structures (pp. 37, 153). In *Cantharellus* (p. 123) the spore-producing layer is ridged and grooved.

AGARICALES. All the species in this group have mushroom-shaped spore-producing structures — a type of reproductive body called a pileus. The group includes *Boletus* (pp. 143, 109), in which the spore-producing layer develops inside tubes, and the large number of genera in which the spores are produced on the surface of gills. Of these, *Coprinus* (pp. 35, 41), *Psathyrella* (p. 139), and *Panaeolus* (p. 41) have black spores; *Agaricus* (pp. 31, 103), *Stropharia* (p. 31), *Psilocybe* (p. 33), and *Hypholoma* (p. 141) have purple spores; *Nolanea* (p. 31, 103) and *Pluteus* (p. 135) have pink spores; and the other fifteen genera mentioned in this book have white spores.

GASTEROMYCETALES. In this group the spores are produced within the fungus rather than in an exposed layer (p. 155).

TREMELLALES. These fungi have a jelly-like texture (p. 149).

Ascomycetes

These are characterised by spores produced, usually in groups of eight, within tiny, elongated, oval sacs called asci. When ripe, they are discharged violently for a considerable distance — usually at least a hundred times as great as far as are the spores of the Basidiomycetes. Consequently, asci are never found inside tubes or on the surface of gills, but are usually produced in horizontal layers from which the spores are shot upwards. The Ascomycetes described in this book are classified into four groups.

PEZIZALES. In this group the spore-producing layer develops as the lining of a shallow cup (p. 150). The asci have small lids at their tips which come off when the spores are discharged.

TUBERALES. The spores are produced within the fungus rather than in an exposed layer, and the asci do not discharge their spores violently. The example given in this book is *Tuber* (p. 155).

HELOTIALES. In this group the spore-producing layers develop in cups similar to those of PEZIZALES, but the asci open by pores and do not have lids. The genera mentioned here are *Calycella* and *Helotium* (p. 145), *Bulgaria* and *Coryne* (p. 147), and *Chlorociboria* (p. 151).

SPHAERIALES. In this group the spores are produced in minute, flask-shaped cavities. Each ascus, as it ripens, protrudes through the opening at the top of the 'flask' and discharges its spores. The genera mentioned here are *Poronia* (p. 41), *Hypoxylon*, *Diatrype*, and *Nectria* (p. 145), and *Xylosphaeria*, *Ustulina*, and *Daldinia* (p. 147).

ALGAE

This is a very large and successful group of plants which flourish in the sea, in fresh water, and in damp places on land. They have a great variety of structure. Many of them are only single celled, and so individually are invisible except with a microscope, although in a mass they may be very conspicuous. Some consist of fine threads or filaments, which may be branched or unbranched. Others develop into quite large and elaborate plants, as do many seaweeds. In this book the many species that grow in fresh water are omitted. The algae are divided into four main groups, largely on a basis of the colours of the pigments they contain.

Chlorophyceae

These are grass-green in colour and contain the same pigments, including chlorophyll, as do higher plants. Many, including *Pleurococcus* (p. 173), consist of single cells, and species of a related genus, *Trebouxia*, often form the algal partner in lichens (p. 66). Examples of filaments consisting of rows of cells are *Hormidium* (p. 173) and *Cladophora* (p. 15); while *Ulva* (p. 9), *Monostroma* (p. 9), and *Enteromorpha* (p. 27) also have a cellular structure, but grow in the form of sheets or tubes. *Codium* (p. 15) is built up from fine, branching threads without cross walls.

Phaeophyceae

In this group the presence of chlorophyll is hidden by a brown pigment, fucoxanthin. If a plant is plunged in hot water, the pigment dissolves, and the frond becomes bright green. Most

Phaeophyceae are marine and flourish in the colder seas of the world. No one-celled forms are known. Most of the species described in this book belong to three main groups.

ECTOCARPALES. This group includes filamentous species such as *Ectocarpus* and *Pilayella* (p. 13), and others with fronds of a jelly-like texture, for example *Eudesme*, *Leathesia*, and *Chordaria* (p. 17).

FUCALES. The conspicuous and common seaweeds from between tide marks belong to this group: *Pelvetia* (p. 3), *Fucus*, *Ascophyllum*, and *Himanthalia* (p. 5), *Bifurcaria* and *Cystoseira* (p. 13), and *Halidrys* (p. 12).

LAMINARIALES. This group includes the largest of all British seaweeds: *Alaria*, *Laminaria*, *Saccorhiza*, and *Chorda* (p. 19).

Rhodophyceae

These contain a red pigment, phycoerythrin, as well as chlorophyll. They are nearly all marine, often growing in deep water, and they are especially abundant in tropical seas. Representatives from four large groups are described in this book.

GIGARTINALES. The plants in this group mostly have flattened, branching fronds, for example, *Gigartina* and *Chondrus* (p. 9), and *Calliblepharis* (p. 21).

CRYPTONEMALES. Many species in this group become encrusted with a deposit of calcium carbonate, for instance, *Corallina* and *Jania* (p. 15), which are filamentous. *Lithothamnion* and *Lithophyllum* (p. 7) form hard crusts on rock and might be mistaken at a first glance for encrusting lichens.

RHODYMENIALES. Many of the tropical members of this group have large, hollow fronds, and in Britain, *Lomentaria* and *Gastroclonium* (p. 11), and *Chylocladia* (p. 15) are to some extent hollow. *Rhodymenia* (p. 19) is flat and not hollow.

CERAMIALES. These are mostly filamentous forms, for instance, *Ceramium* (p. 15) *Griffithsia* (p. 11), and *Polysiphonia* (p. 5).

Cyanophyceae

The plants in this group are very simply constructed and contain a blue pigment, phycocyanin, as well as chlorophyll. All of them are either single celled or consist of unbranched filaments. Unlike all other plants, except some bacteria, they are able to absorb and utilize nitrogen from the atmosphere, and are thus able to grow in waters and on moist soil where nitrogen-containing compounds are scarce. Species of *Nostoc* often form the algal partner in lichens. Two filamentous species, *Calothrix* and *Rivularia*, are mentioned on p. 26.

OTHER GROUPS

Lichens

Lichens are dual organisms, consisting of a fungus, which gives the plant its shape and form, living in very close association with an alga, which, like other green plants, can use the light energy of the sun to manufacture sugar and other food substances from simple materials. This close association of two very different kinds of plant is able to produce a more elaborate and longer-lived structure than either could form alone (p. 66). The algal partner is either a green alga, frequently a species of *Trebouxia*, or a blue-green alga, usually *Nostoc*, and reproduces by simple division of its cells. The fungus partner reproduces by means of spores. In many lichens, special powdery reproductive structures (soredia) are also produced, each 'granule' of the powder consisting of a few algal cells surrounded by fungus threads. In a few tropical lichens the fungus is a member of the group Basidiomycetes (p. 195), and this may also be so in *Coriscium* (p. 85).

In most lichens the fungus is a member of the group Ascomycetes (p. 196). Those in which the spores develop on the inner surfaces of shallow cups are members of the PEZIZALES or HELOTIALES. Others, in which the asci develop in flask-shaped cavities, are members of the SPHAERIALES. About 150 genera of lichens are found in Britain.

Myxomycetes

The English name ('slime moulds'); sometimes used for this group, is not at all apt, for they are not made up of fine threads (hyphae) and are not at all closely related to fungi. When actively growing, they consist of a mass of living 'jelly' or protoplasm, with no fixed shape, which creeps slowly about on the surface of rotten wood or other decaying plant material. They obtain their food by digesting vegetable matter, and may frequently be found feeding on fungi. When fully grown, they cease moving, and the protoplasm becomes converted into large numbers of spore cases. These may grow separately on stalks, as in *Stemonitis*, or they may be joined together in masses (aethelia), as in *Lycogala* (p. 173). About fifty genera of Myxomycetes are found in Britain.

Charophyta

The members of this group are sometimes called 'Stoneworts' because they frequently become encrusted with a deposit of calcium carbonate (p. 92). They are a small and ancient group of plants which are common and are widely distributed in fresh water throughout the world. A remarkable feature of *Nitella* is that each section of stem between one whorl of branches and the next is a single cell. In *Chara* there is a similar very long cell in each section of main stem, but this is not obvious because it is covered with a layer of smaller cells. The reproductive structures are also unique (*see Chara*, p. 92). Their nearest relatives appear to be green algae (Chlorophyceae), but in view of their special features, the relationship cannot be very close.

ASSOCIATIONS BETWEEN FLOWERLESS PLANTS

The plants described in this book are grouped as far as possible in the places where they most commonly grow. In these habitats they form part of a community of plants and animals in which every species affects to some extent the life of every other. The animals, many of which are very small, affect the plants especially by feeding on them, and the plants affect the animals mainly by serving as food and by providing protection from the weather and from other animals. Sometimes the association between a particular animal and a particular plant is especially close; for example, the tube worm *Spirorbis borealis* is usually found on the seaweed *Fucus serratus* (p. 5), and the tiny animal *Callidina symbiotica* lives in the pitchers of the liverwort *Frullania dilatata* (p. 176).

The plants affect one another considerably by competing for light and water, but they also provide mutual shelter. Often one of the species in the community has an especially marked influence, and is then said to be dominant. For instance, whether or not a shade-loving plant such as the fern *Dryopteris aemula* (p. 185) can grow in a particular woodland depends on the degree of protection provided by the dominant trees. On the other hand, in grassland the dominant grasses prevent the growth of all but the most vigorous mosses, such as *Pseudoscleropodium purum* (p. 43). In most habitats it is a flowering plant that is dominant, but there are some in which flowerless plants form the most conspicuous or the only vegetation. They are often colonizers of newly-formed habitats, such as bonfire sites (p. 41), or landslides, screes, and bare rock surfaces. In such situations the growth of the dominant spore plant affects local conditions and has a considerable influence on the other plants that are able to grow in the habitat. For instance, the bog mosses of the genus *Sphagnum* (pp. 86, 88) absorb and retain large quantities of water and encourage wet and acid conditions in which a limited number of rather specialized species are able to grow. Such habitat associations are similar to those that develop amongst flowering plants, but in addition to these, spore plants are often involved in much closer associations, some of which are difficult to explain. An example is that of the lichen *Normandina pulchella* and the liverwort *Frullania dilatata* (p. 177), which are very often found growing together.

On the seashore, almost the whole of the vegetation is made up of seaweeds (algae). The various species grown in zones at different levels on the beach, each zone being occupied by a characteristic community of plants and animals. The position of a particular zone in relation to the tide marks, and its extent and width, depend on the steepness of the slope and the degree of exposure of the shore. The barnacle zone provides a useful reference line (p. 6). Above it, *Pelvetia canaliculata* (p. 3) is often the dominant seaweed. Below it is a broad zone with species of *Fucus* (p. 5), and farther down the beach at and below low water mark is a *Laminaria* zone (p. 19).

Lichens and mosses are the only plants that can gain a foothold on bare rocks, both by the sea and in upland areas. These pioneer plants gradually break down the rock surface into fine particles which, mixed with decaying parts of the plants themselves, form soil in which other plants may grow. The lichen *Rhizocarpon geographicum* (p. 71), for example, and moss species of the genus *Andreaea* (p. 61) are found everywhere at high altitudes on hard rocks other than limestone. Many species of crust-forming lichens, for instance *Lecanora calcarea* (p. 77), are abundant on limestone, and the moss *Tortella tortuosa* (p. 61) forms large cushions which gradually etch deep pits in this type of rock.

The other main kind of habitat in which flowerless plants are frequently the dominant species is on the bark of trees. Mosses, liverworts, and lichens are often abundant, growing as epiphytes; they are merely attached to the tree, and do not obtain food from it as parasitic species do. In very damp woodlands lichen species of the genus *Sticta* (p. 167) may be dominant. In woods where the atmosphere is not quite so moist, their place is taken by species of *Parmelia* (p. 163) and *Physcia*

(p. 165), and in drier places lichens which form crusts are dominant. Some, such as species of *Graphis* (p. 171), are restricted to smooth bark, while others, for instance species of *Pertusaria* (p. 169), are commonest on rough bark. All these plants obtain the water and mineral salts they require from rain water falling on them directly and running down the tree trunks, which they are able to absorb over the whole of their surface. For this reason they are very sensitive to poisonous substances present in the polluted air of towns, especially sulphur dioxide. The few species that are able to tolerate atmospheric pollution are shown on p. 173. Epiphytic mosses and liverworts are illustrated on pp. 175 and 177. Epiphytic ferns are abundant in the tropics, but the only British species which grow as epiphytes are species of *Polypodium* (p. 189). A number of smaller species of seaweeds grow as epiphytes on larger species, for instance *Pilayella littoralis* (p. 13) and *Brogniartella byssoides* (p. 25). *Polysiphonia lanuginosa* (p. 5) is an interesting example for, unlike the vast majority of epiphytes, it is specific, that is, it is rarely found except on *Ascophyllum nodosum*. Also it often has another smaller red seaweed growing epiphytically on it.

Many fungi are parasitic, that is, they grow on and obtain their food from another living plant. Some attack a variety of hosts, but others are specific, growing on one species only. A good example is *Piptoporus betulinus* (p. 117) which is found only on birch. *Fomes fomentarius* (p. 117) is also restricted to birch in Scotland, but it grows on beech in one place in southern England.

Fungi that grow on the ground obtain their food from decaying plant and animal remains, and are called saprophytes. Their actively growing and feeding parts are branching wefts of fine threads (hyphae) spreading through the soil. In many woodland species some of the hyphae enter into a close association with the roots of trees, penetrating them to some extent and forming a dense felted covering on the surface. The roots become modified, and instead of being slender and fibrous with long branches coming off at narrow angles, like the ordinary water-absorbing roots, they are thick and fleshy, with short thick side branches developing at right angles. The modified root with its covering of fungus hyphae is called a mycorrhiza. Some species of mycorrhizal fungus form a specific relationship with one kind of tree, for instance *Amanita muscaria* (p. 113) with birch and *Boletus elegans* (p. 109) with larch, but many others are commonly associated with more than one kind. The tree certainly benefits from the association, for water and mineral salts absorbed from the soil by the fungus are rapidly transferred to the roots. The fungus probably obtains food materials such as sugar from the roots in return, and so the relationship is a case of symbiosis in which both partners benefit.

In other plants that form mycorrhizas, such as orchids, the association is one of controlled parasitism. The most striking example is the relationship between the fungus *Armillaria mellea* (p. 141) and the orchid *Gastrodia elata*. *A. mellea* is a destructive parasite of trees found commonly in forests in temperate regions throughout the world. In Japan, the hyphae sometimes penetrate the dormant root tubers of *G. elata*. The orchid is stimulated to grow, producing flowers and seeds, and in doing so it obtains its food supply entirely from the fungus. Thus *A. mellea* attacks as a parasite but is, as it were, defeated and enslaved. The relationship between the alga *Pelvetia canaliculata* and the fungus *Mycosphaerella pelvetiae* (p. 2) is probably another rather different example of controlled parasitism. The fungus is always present in the alga, which does not appear to be affected in any way.

A somewhat similar delicate balance between two very different kinds of plant is found in lichens, each species of which is a close association between a fungus and an alga (p. 66), but here both partners benefit, and the relationship is called symbiosis. The simplest of all lichens in *Botrydina vulgaris* (*see Omphalina sphagnicola*, p. 88), and another especially interesting association is that between *Coriscium viride* and *Omphalina ericitorum* (p. 85). In many lichens, for instance *Peltigera canina* (p. 29), the algal partner is a species of the blue-green alga *Nostoc*. This plant is able to absorb and utilize nitrogen from the atmosphere, and is thus able to make a special contribution to the nutrition of the lichen. *Nostoc* is also found closely associated with a number of other plants, for instance, the liverwort, *Blasia pusilla* (p. 101). A related genus, *Anaboena*, is associated with the fern *Azolla filiculoides* (p. 96). In these cases the basis of the association can be at least partly explained, but there are many others which are not yet fully understood.

The many different kinds of association that exist between flowerless plants is a fascinating and complex subject, to which this is but a simple introduction.

SUGGESTIONS FOR FURTHER READING

Readers who wish to carry their study of flowerless plants further, especially in matters of growth and development and the microscopic details of reproduction and spore production, are recommended to look in one of the many general text books of botany that are available. Especially recommended is *Flowerless Plants*, by D. H. Scott and C. T. Ingold, Black.

The following books are listed because they should be of help to those who wish to carry their study of any group further. The best descriptions of British Ferns that are available are to be found in *Welsh Ferns* by H. A. Hyde and A. E. Wade, Museum of Wales, which, in spite of its title, deals with all the ferns that grow in Britain.

Illustrated descriptions of Clubmosses and Horsetails are included in *British Ferns and Mosses* by P. G. Taylor, 1960, Eyre and Spottiswoode. The descriptions of Horsetails, especially, are excellent.

A useful book on the Bryophyta for beginners is *British Mosses and Liverworts*, by E. V. Watson, C.U.P.

Of the many handbooks on fungi that are available, *A Guide to Mushrooms and Toadstools*, by M. Lange and F. B. Hora, Collins, is to be recommended. In spite of its title, it deals with all the larger fungi. *The Observers' Book of Common Fungi* by E. M. Wakefield, Warne, is also useful. There is no handbook on British Myxomycetes at present in print, and the only one available on lichens is *The Observers' Book of Lichens* by K. L. Alvin and K. A. Kershaw, 1963, Warne.

The standard work on Charophyta is *British Stoneworts* by G. O. Allen, Haslemere Museum, and *British Seaweeds* by C. Dickinson, Eyre and Spottiswoode, is a useful book on marine Algae.

Readers may come across unfamiliar scientific names in older books, but their up-to-date equivalents can be found out by looking in one of the following standard Check Lists:—

A Revised Preliminary Census List of British Pteridophytes. A. C. Jermy. 1960.

An Annotated List of British Mosses. P. W. Richards and E. C. Wallace. 1950.

An Annotated List of British Hepatics. E. W. Jones. 1958.

New Check List of British Agarics and Boleti. R. W. G. Dennis, P. D. Orton and F. B. Hora. 1960.

A Monograph of Clavaria and Allied Genera. E. J. H. Corner. 1950

Atlas des Champignons de l'Europe; Polyporaceae. A. Pilat. 1936.

British Cup Fungi. R. W. G. Dennis. 1960.

Guide to the British Mycetozoa. G. Lister. 1919.

A Revision of the Characeae. R. D. Wood and K. Imahori. 1964.

A New Check List of British Lichens. P. W. James. 1965.

A Revised Check List of British Marine Algae. M. Parke and P. S. Dixon. 1964.

INDEX

The numbers in heavy type refer to pictures

DATE DUE